LOCAL ECONOMIC DEVELOPMENT AND THE ENVIRONMENT

Concerns about the impact of economic development upon the global environment have increased in recent years. Sustainable development has been proposed as a means of reconciling the pressures between the two, allowing the integration of economic, environmental and social concerns. While policy makers at international, national and local levels have rapidly adopted sustainable development as a key aim, there is a great deal of uncertainty as to what it means in practice.

This book focuses upon the potential to integrate economic and environmental policies at the local and regional scale. Local initiatives are investigated within their wider economic and environmental policy contexts in order to illustrate both the constraints and opportunities for local policy makers. Attention is given to global economic trends, as well as to the specific policy contexts of the European Union and the national contexts of the UK, USA, Australia, Japan and Sweden. The key principles for designing integrative policies and descriptions of initiatives and projects in a variety of locations are also considered.

This book will be of vital interest not only to students and academics working in the field of local and regional economic development, but also to policy makers and planners seeking guidelines for the role of local government in the local economy, sustainable development and the implementation of policies.

David Gibbs is Professor of Human Geography at the University of Hull. His research interests are in the field of local and regional economic development, with a paticular focus upon the implications of both technological change and environmental issues for localities and regions. He is the author of several articles on these topics in a number of international geographical journals.

ROUTLEDGE RESEARCH GLOBAL ENVIRONMENTAL CHANGE SERIES

Editorial Advisory Board: Alister Scott, Jim Skea, Michael Redclift, Martin Parry, Timothy O'Riordan, Robin Grove-White, Kathy Homewood, Judith Rees, Camilla Toulmin, Kerry Turner, Richard Welford and Stephen Yearley

1 THE CONDITION OF SUSTAINABILITY
Ian Drummond and Terry Marsden

2 WORKING THE SAHEL
Environment and society in Northern Nigeria
Michael Mortimore and Bill Adams

3 GLOBAL TRADE AND GLOBAL SOCIAL ISSUES
Edited by Annie Taylor and Caroline Thomas

4 ENVIRONMENTAL POLICIES AND NGO INFLUENCE
Land degradation and sustainable resource management in Sub-Saharan Africa
Edited by Alan Thomas, Susan Carr and David Humphreys

5 THE SOCIOLOGY OF ENERGY, BUILDINGS AND THE ENVIRONMENT
Constructing knowledge, designing practice
Simon Guy and Elizabeth Shove

6 LIVING WITH ENVIRONMENTAL CHANGE
Social vulnerability, adaptation and resilience in Vietnam
Edited by W. Neil Adger, P. Mick Kelly and Nguyen Huu Ninh

7 LAND AND LIMITS
Interpreting sustainablity in the planning process
Susan Owens and Richard Cowell

8   THE BUSINESS OF GREENING
    Edited by Stephen Fineman

9   INDUSTRY AND ENVIRONMENT IN LATIN
    AMERICA
    Edited by Rhys Jenkins

10 LOCAL ECONOMIC DEVELOPMENT
    AND THE ENVIRONMENT
    David Gibbs

Also available in the Routledge Global Environmental Change Series:

TIMESCAPES OF MODERNITY
Barbara Adam

REFRAMING DEFORESTATION
James Fairhead and Melissa Leach

BRITISH ENVIRONMENTAL POLICY
AND EUROPE
Edited by Philip Lowe

THE POLITICS OF SUSTAINABLE
DEVELOPMENT
*Edited by Susan Baker, Maria Kousis, Dick Richardson
and Stephen Young*

ARGUMENT IN THE GREENHOUSE
Mick Mabey, Stephen Hall, Clare Smith and Sujata Gupta

ENVIRONMENTALISM AND THE
MASS MEDIA
*Graham Chapman, Keval Kumar, Caroline Fraser
and Ivor Gaber*

ENVIRONMENTAL CHANGE IN
SOUTHEAST ASIA
Edited by Michael Parnwell and Raymond Bryant

THE POLITICS OF CLIMATE CHANGE
Edited by Timothy O'Riordan and Jill Jagger

POPULATION AND FOOD
Tim Dyson

THE ENVIRONMENT AND
INTERNATIONAL RELATIONS
Edited by John Vogler and Mark Imber

GLOBAL WARMING AND ENERGY DEMAND
Edited by Terry Barker, Paul Ekins and Nick Johnstone

SOCIAL THEORY AND THE
GLOBAL ENVIRONMENT
Michael Redclift and Ted Benton

LOCAL ECONOMIC DEVELOPMENT AND THE ENVIRONMENT

David Gibbs

Routledge
Taylor & Francis Group
LONDON AND NEW YORK

First published 2002
by Routledge
11 New Fetter Lane, London EC4P 4EE

Simultaneously published in the USA and Canada
by Routledge
29 West 35th Street, New York, NY 10001

Transferred to Digital Printing 2003

Routledge is an imprint of the Taylor & Francis Group

© 2002 David Gibbs

Typeset in Sabon by Exe Valley Dataset Ltd, Exeter
Printed and bound in Great Britain by
TJI Digital, Padstow, Cornwall

All rights reserved. No part of this book may be reprinted or reproduced or utilised in any form or by any electronic, mechanical, or other means, now known or hereafter invented, including photocopying and recording, or in any information storage or retrieval system, without permission in writing from the publishers.

British Library Cataloguing in Publication Data
A catalogue record for this book is available
from the British Library

Library of Congress Cataloging in Publication Data
Gibbs, David, 1955–
Local economic development and the environment / David Gibbs.
p. cm.
Includes bibliographical references and index.
1. Sustainable development. 2. Economic development–Environmental aspects. I. Title.

HD75.6.G52 2002
658.8′3–dc21   2001048176

ISBN 0–415–16825–2

FOR JANET

CONTENTS

List of tables — xi
List of boxes — xii
Preface — xiii
Acknowledgements — xv

1 Approaching local sustainability — 1

Introduction 1
Sustainable development 2
Ecological modernisation 7
Urban regime theory 12
Regulation theory 15
Conclusions 23

2 The changing context for local economic development — 26

Introduction 26
Changing international and national contexts 27
Conclusions 43

3 The changing environmental policy context for local action — 51

Introduction 51
The international policy context 52
Differing national contexts 65
Conclusions 82

4 Integrating economic development and environmental strategies at the local level — 85

Introduction 85
*Principles for sustainable development
  in local areas and regions* 88

CONTENTS

*Principles for sustainable local and
regional economies 93
Implementing sustainable local and regional
economic development: bioregionalism and
regional environmental management systems 100
Conclusions 106*

5 **Sustainability and economic regeneration: making it
happen on the ground** 109

*Introduction 109
Land use and transport planning 110
Advice and support for local business
to improve environmental performance 111
Using technology for sustainability 113
Promoting green consumerism and purchasing
policies 115
Targeted inward investment strategy 116
Changing governance structures 117
Promoting the environmental business sector 120
Linking economic development, labour markets
and social policy 122
Partnerships between local government and industry 125
Provision of sustainability infrastructure 127
Industrial ecology and eco-industrial parks 128
Indicators and output measurement 133
Conclusions 135*

6 **Integrating economic development and the environment:
future prospects for local areas** 139

*Introduction 139
Reflections on theory 140
The strengths and weaknesses of a local approach 144*

References 151
Index 168

TABLES

1.1	The spectrum of sustainable development	4
1.2	Characteristics of 'weak' and 'strong' ecological modernisation	9
1.3	Phases of regulation and accumulation	16
1.4	Environmental issues and possible regulatory solutions	24
2.1	Local strategies in the United States	45
4.1	Guiding principles for sustainable urban development	89
5.1	Factors involved in developing an industrial ecology approach to industrial estates	130
5.2	Criteria for sustainability appraisal of economic development in Cumbria	134

BOXES

1.1	Modes of social regulation	21
2.1	Impacts of international trade on the environment	34
3.1	Changing approaches to the environment in the Maastricht Treaty	56
3.2	Key tasks for the European Environment Agency	57
3.3	Key areas in the EU's Fifth Environmental Action Programme for local authorities	58
3.4	The ISIS Project: an example of LIFE funding	62
3.5	Japanese policy initiatives on the environment	72
3.6	Barriers to better environmental management in Australian local government	75
3.7	Components of a Local Agenda 21 process	79
4.1	Characteristics of a 'sustainable society which promotes economic success'	94
4.2	Strategy to develop a Regional Environmental Management System	104
5.1	Activities in Turin Environment Park	123

PREFACE

My initial interest in the environmental implications of local and regional economic development came about through a commissioned piece of research for the Centre for Local Economic Strategies in Manchester back in 1992. Since that time, my interest has further developed through conducting a number of research projects in this area, in the form of academic projects funded through the UK research councils, as well as more consultancy-type work for practitioners. This book represents an attempt to pull much of this work together into some sort of coherent whole. Coming from a background in economic geography, I have noticed a general lack of interest in the environmental consequences of economic development by most other economic geographers. Similarly, though, when reading much of the environmental literature, and the proposals that are made for sustainable development and more environmentally-aware policies, I have also noticed a lack of awareness of the broader economic context for such initiatives. In both cases I think this is a pity, as both sets of literature and authors have much to offer each other. Obviously there are exceptions to the lack of overlap and some researchers *have* drawn upon both literatures, but I think there remains considerable scope for much more work that attempts to combine the two interpretations. The importance of doing so relates, in part, to the much greater attention paid to environmental issues in economic policy at all spatial scales in recent years. Obviously, this also relates to the serious threat that such ecological problems pose to economic and social activities as they currently exist. Although not everyone is convinced of the magnitude of such changes, it is widely agreed that ecological problems such as enhanced global warming, changing weather patterns and sea level rise are occurring and that the current organisation of economic activities is in large part responsible. The concept of sustainable development, widely publicised following the Rio Earth Summit in 1992, suggests that economic, environmental and social aims can now be made compatible. Exactly how this is to be done has remained very unclear, despite the widespread adoption of at least the rhetoric of sustainability in policy documents at international, national and

PREFACE

local scales. It is also notable that many of these policy documents, from Agenda 21 onwards, privilege the local scale as the key site for policy intervention, though frequently without much justification other than arguing that it is somehow the 'most appropriate' level for intervention. This book therefore attempts to explore some of these issues in greater depth. In particular I have been concerned to try and set out the broader international and national contexts within which local and regional attempts to introduce policies and strategies to combine economic development and environmental issues can be situated. Hopefully, this will lead to a greater understanding by both academics and policy makers of both the potential for, and limits to, such integrative policy initiatives.

David Gibbs
June 2001

ACKNOWLEDGEMENTS

This book draws upon two research projects funded by the Economic and Social Research Council (Grant numbers L320253132 and R000237997), together with work funded by the Local Government Association and the Improvement and Development Agency. I am grateful to the Faculty of Arts Research Committee at Monash University for a Collaborative Visiting Scholarship which provided me with the opportunity to complete the final manuscript. A number of individuals have helped to develop my understanding of the issues explored in this book, though they may not have always realised this. My thanks in particular to Graham Haughton, Mike Jacobs, Tim Jenkins, Andy Jonas, Jim Longhurst, Darryn McEvoy, Simon Marvin, John Shutt and Aidan While. Some of the ideas in the book were formulated during various periods of study leave in Australia – thanks to Chris Cocklin, Bob Fagan, John Langdale, Pauline McGuirk and Phil O'Neill for their help and hospitality. My colleagues at Hull University, notably David Atkinson, Sally Eden, Suzy Reimer, David Sibley and Derek Spooner, have provided a highly stimulating and supportive atmosphere which in no small measure has contributed to my own intellectual development over the past five years. Finally, my thanks to Annabel Watson at Routledge for her advice and patience in answering my queries.

Permission to reprint copyright material has been granted by the following. Table 1.3 reprinted from *Progress in Human Geography*, Volume 16, Tickell, A and Peck, J 'Accumulation, regulation and the geographies of post-Fordism: missing links in regulationist research', 190–218 (1992) with permission from Arnold Publishers. Box 1.1 compiled from *Geoforum*, Volume 23, Peck, J and Tickell, A, 'Local modes of social regulation? Regulation theory, Thatcherism and uneven development', 347–363, Copyright (1992) with permission from Elsevier Science. Box 2.1 from Welford, R (1997) *Hijacking Environmentalism: Corporate Responses to Sustainable Development* with permission from Earthscan Publications Ltd. Box 3.7 and Box 4.1 from Department of the Environment, Transport and the Regions (1997) *Sustainable Local*

ACKNOWLEDGEMENTS

Communities for the 21st Century and Table 4.1 from Haughton, G and Hunter, C (1994) *Sustainable Cities* with permission from Her Majesty's Stationery Office. Box 4.2 from Welford, R (1995) *Environmental Strategy and Sustainable Development: The Corporate Challenge for the 21st Century*, with permission from Routledge. Box 5.1 with permission from Davide Domosso of Turin Environment Park. Table 5.2 from Forum for the Future (1998) *Case Studies of Sustainable Local Economic Development* with permission from Forum for the Future.

1

APPROACHING LOCAL SUSTAINABILITY

Introduction

In recent years the issue of environmental change has become a key area of debate. There is a widespread concern that the consequences of industrialisation are increasingly negative and that action needs to be taken to remedy this. Although there is no absolute consensus that major environmental changes are occurring (see, for example, the formidable pressures brought to bear on the US President in the run up to the Kyoto climate change summit in 1997 by representatives of the automobile, steel and oil industries), there is a broad agreement that these changes are in train and that some form of response is needed. As Dryzek states:

> Today, any credible political–economic vision must address the challenge presented by ecological problems. 'The environment' can no longer be thought of as just one issue among many. Ecological problems are sufficiently widespread and serious to constitute an acid test for all actual and proposed political and economic arrangements, be they incremental or revolutionary.
> (Dryzek, 1994: 176)

Where this agreement breaks down is in the form of the appropriate response to environmental change. There is a wide diversity of opinion here from deep green ecologists who require a wholesale restructuring of society, through to some economists who believe that market instruments are capable of restoring the 'environmental equilibrium' and that the basic socio-economic form can remain intact (Torgerson, 1995). While there is a need for an international and national response to environmental problems, the aim of this book is more modest than setting out a global plan of action. This book is concerned with the potential for, and form of, action at the scale of local and regional economies. In particular, it examines the potential for the integration of economic and environmental policies at this scale. The rationale for this is partly related to the major emphasis placed upon the local as the site for the delivery of environmental policy and sustainable development (see, for example, the United Nation's Agenda 21 programme (United Nations, 1993) and the European Union's Fifth Environmental Action Programme (Commission of the European Communities, 1992)), but also because I believe that the local scale has some level of causal autonomy which necessitates action at this scale, albeit located within an international and national context (Haughton and Hunter, 1994). As Churchill and Worthington (1995: 98) state 'to see only the local is to be reactive; to see only the global can be abstract and irrelevant'.

This chapter provides an outline of the theoretical background which informs this approach to the study of the local economy and the environment. In particular, the argument is that there is merit in drawing upon a combination of research which has attempted to contextualise environmental change (sustainable development and ecological modernisation), together with recent developments within political economy. The latter has recently focused upon two theoretical arguments – urban regime theory and regulation theory. Even though the differences between these two literatures are significant (see the debates within Lauria, 1997), useful insights can be gained from a combination of these two approaches. To date the theoretical underpinnings of environmental policy and sustainability have been weakly developed and this, I believe, has hindered the development of more concrete policy interventions (Nijkamp and Perrels, 1994). As Leff states:

> In the acritical discourse of sustainable development and "natural capital", environmental issues are considered part of the "new economic order" and of the global transition to liberal democracy. Although these issues have generated a social response, visible in the new environmental movements, the latter still lack a theoretical framework and a strategic programme for the construction of an ecologically sustainable productive rationality.
>
> (Leff, 1996: 138)

It is hoped that by attempting to set environmental issues within a more developed theoretical framework, such a conceptualisation will not only allow a clearer understanding of the processes at work, but also, more pragmatically, enable the development and implementation of practical policy initiatives at the local scale.

Sustainable development

Although the phrase was used before this date, the term sustainable development was popularized by the publication of the Brundtland Report in the late 1980s (World Commission on Environment and Development, 1987). The definition used by Brundtland emphasised meeting the needs of the present without compromising the ability of future generations to meet their own needs. In a consultation paper for the UK's revised sustainable development strategy, the Department of the Environment, Transport and the Regions (1997b: 4) define it as 'ensuring a better quality of life for everyone, now and for generations to come' and based on four objectives:

- social progress which recognises the needs of everyone;
- effective protection of the environment;
- prudent use of natural resources;
- maintenance of high and stable levels of economic growth.

While definitions of sustainable development vary, most allude to the following core principles:

- quality of life (including and linking social, economic and environmental aspects);
- care for the environment;
- thought for the future and the precautionary principle;
- fairness and equity;
- participation and partnership.

The United Nations Earth Summit, held in Rio de Janeiro in 1992, seemed to have placed the notion of sustainable development firmly on the policy making agenda. As Healey comments:

> the new substantive agenda is now focused in theory around the relation between environmental quality (understood in biospheric as well as aesthetic and utilitarian terms) and economic development, expressed in the rhetoric of *sustainable development*. Economic development is understood not merely as the promotion of growth, nor narrowly as property development, but as the promotion of the assets of area.
> (Healey, 1995: 263)

A variety of different perspectives have been used to approach the concept of sustainable development (Pezzey, 1992; Turner, 1993). A spectrum of perspectives can be identified ranging from a technocentric 'very weak sustainability' position through to an ecocentric position of 'very strong sustainability' (see Table 1.1). This spectrum from weak(er) to strong(er) versions of sustainability is important because the way in which sustainable development is defined and operationalised crucially shapes how the economy and the environment are integrated. Advocates of weak sustainability approaches assume that there is a very high degree of substitutability between human capital and natural capital (Pearce et al.,

Table 1.1 The spectrum of sustainable development

Version	Features
Very weak sustainability	Overall stock of capital assets remains stable over time, complete substitution between human and natural capital. Essential link between willingness to pay and sustainable development.
Weak sustainability	Limits set on natural capital usage. Some natural capital is critical i.e. non-substitutable. Related to the precautionary principle or safe minimum standards. Tradeoffs still possible.
Strong sustainability	Not all ecosystem functions and services can be adequately valued economically. Uncertainty means whatever the social benefits forgone, losses of critical natural capital are not possible.
Very strong sustainability	Steady-state economic system based on thermodynamic limits and constraints. Matter and energy throughput should be minimised.

Source: Adapted from Turner (1993).

1994). In these approaches environmental concerns assume a higher priority in economic policy, but there is no specification of the environmental quality to be achieved. The emphasis will effectively be on raising environmental efficiency, i.e. reducing the environmental impact of each unit of economic activity and addressing individual parts of the economy, such as firms or sectors, without an holistic approach to the environment. Strong versions of sustainability, however, take issue with the assumption of almost infinite substitutability of resources and specify minimum levels of environmental quality to be achieved prior to consideration of other goals (Turner, 1993). Strong versions of sustainable development begin from a presumption that society cannot simply let economic activity result in a continual decline in the quality and functions of the environment, even though it may be beneficial in other ways (Daly and Cobb, 1989; Jacobs and Stott, 1992).

In strong versions of sustainable development there are lower limits to environmental quality, such that a sustainable economy is a constrained economy (Jacobs, 1991). Strong versions of sustainable development will require targets to be set for environmental impacts, such as emission levels, and measures taken to constrain firms and individuals to ensure that these targets are met such that the whole of the economy is affected, rather than some of its constituent parts (Jacobs and Stott, 1992). The spectrum is thus associated with differing views on the required degree of intervention into the economic system, which is necessary and sufficient for sustainable development (Turner, 1993). Under weak versions of sustainability the market and the preservation of the status quo play a much greater role than in strong versions, where direct intervention is necessary and where greater social and institutional change is required (O'Riordan and Jordan, 1995).

These notions of weak and strong sustainability have also been applied in relation to the strength of integrative activity within environmental policy (Gouldson and Murphy, 1996). Strong integration exists where departments within organisations undergo a process of internal culture change, involving modification of strategic and operational characteristics. At the opposite end of the spectrum is weak integration, involving the adjustment of existing activities within existing operational boundaries. Gouldson and Murphy (1996) query whether weak integration is compatible in the longer-term with the holistic concepts of sustainable development, as the former involves the adjustment of existing structures and policies rather than the types of institutional innovation and policy change necessitated by the latter. Ravetz (1996) outlines a spectrum of sustainability for urban areas. At the lowest level the 'sustainable city' is concerned with surface appearances, such as land reclamation and tree planting. The next level focuses upon less tangible or visible problems in the local environment, such as air quality and waste disposal. A third stage looks at the functioning of the urban system as a total metabolism, with internal and external effects. The 'system' is taken to include all aspects of urban infrastructure which predetermine the patterns and impacts of production and consumption (Ravetz, 1996). A further stage still includes all human activity within the urban system, such as production, consumption and indirect linkages such as overseas trading. These notions of a sustainability spectrum are useful as a method to evaluate the degree of commitment to sustainability.

However, there has also been much criticism of the concept of sustainable development and whether it has any practical meaning. As Myerson and Rydin comment:

> does the discourse around the concept of 'sustainable development' represent a cultural oxymoron, a conflation of policy goals from the distinct economic and environmental policy arenas, or is it an innovative step forward in policy thinking which provides new opportunities for goal achievement?
> (Myerson and Rydin, 1994: 439)

One response has been to see sustainable development as simply one discourse of environmentalism amongst several others, albeit increasingly the dominant one (Dryzek, 1997). Similarly, Torgerson (1995) comments on the ambiguity of the term and the way in which this allows political actors from many different backgrounds to proceed without having to agree on what action to take – a benefit to those who see the need for incremental reform rather than radical social transformations of the type advocated by 'deep greens'. Dryzek (1997: 125) argues that this is not surprising given that sustainable development 'is a discourse rather than a concept which can or should be defined with any precision'.

In spatial terms a key component of sustainable development is the adaptation of the old radical green slogan 'think globally, act locally'. Sustainable development is frequently predicated upon the basis of simultaneously shifting some political power *up* to transnational levels of political organisation and *down* to the local scale (Dryzek, 1997). The concomitant of these power shifts is a decrease in the capacity rooted in the nation-state – in part a recognition by environmentalists of the 'hollowing out' of the nation-state and associated globalisation tendencies (Jessop, 1994). At the local scale there is a commitment to exploratory and decentralised approaches to sustainability, with a range of local experimentation (Lee, 1993). Dryzek (1997) sees this as a potential problem (how can these local experiments be harnessed together?) and as an advantage (a welcome antidote to nation-states under the sway of market-led economic approaches). It can be argued that sustainable development is thus a discourse of civil society and not nation-states:

> Sustainable development discourses can be reread as a new power/ knowledge formation, aiming at accumulating power for comparatively powerless subnational and supranational agencies through the mobilisation of new knowledges about the performance of essentially national economies and states that exert their authority to foster development at any cost. Rather than sovereign territories, these discourses look at subnational and transnational domains for sustainable ecosystems . . . to reconfigure the circuits of biopower generation and utilisation.
>
> (Luke, 1995: 29)

For the moment, though, the nation-state remains the locus of most regulatory activity. At this scale, despite the increasing acceptance of sustainable development as the basis for environmental policy at a variety of spatial scales, the trend in recent years in industrialised countries has been for a move towards market-led economies, the globalisation of economic activity and a system of free trade which runs counter to sustainable development (Korten, 1995). Other than amongst 'deep greens', there is a recognition that any future shift in society and the way in which the economy is organised is unlikely to involve radical change, at least in the short to medium term. Over this time-scale, market mechanisms will remain dominant and perhaps the best that can be expected is a gradual shift towards a more sustainable future. Given that a market-based, capitalist economic system looks set to dominate the global economy, one response has been to argue that integrating environmental and economic policy can be both profitable for business and contribute to sustainable development through a programme of ecological modernisation.

Ecological modernisation

The perspective of ecological modernisation is said to offer a constructive approach to deal with environmental problems, with a central role assigned to science and technology (Mol and Spaargaren, 1993). The concept was developed in the 1980s mainly through the initial work of the German social scientists Joseph Huber (1982) and Martin Jänicke (1985). The basic argument is that the central institutions of modern society can be transformed in order to avoid the ecological crisis. Huber (1982, 1985), for example, has argued the need for an 'ecological switchover' – a transition of industrial society towards an ecologically rational organisation of production, based upon the theory of a changed relationship between the economy and ecology. Rather than the deep ecological position of the radical restructuring of society, ecological modernisation has more in common with 'strong' versions of sustainability in that it envisages a process of the progressive modernisation of the institutions of modern society, as opposed to their destruction or dismantling (Mol and Spaargaren, 1993). Both Hajer (1995) and Harvey (1996) link ecological modernisation and sustainable development together such that the latter is the 'central story line' of the policy discourse of ecological modernisation. However, ecological modernisation has much more analytical rigour than sustainable development and 'has a much sharper focus than does sustainable development on exactly what needs to be done with the capitalist political economy, especially within the confines of the developed nation state' (Dryzek, 1997: 143).

As a theoretical framework, the concept of ecological modernisation can be used at two levels. First, it can be used as a theoretical concept to analyse those changes to the central institutions in modern society deemed necessary to solve the ecological crisis. Second, ecological modernisation is used to describe a more pragmatic political programme to redirect environmental policy-making. In the sense of the first meaning, ecological modernisation stands for a major transformation, or 'ecological switch', of the industrialisation process into a direction that takes account of the need to maintain the sustenance base. As with sustainable development, ecological modernisation indicates the possibility of overcoming environmental crises without leaving the path of modernisation (Mol and Spaargaren, 1993). The processes of production and consumption can be restructured on ecological terms through the institutionalisation of ecological aims (Mol, 1994). Three central projects form the heart of this ecological switchover (Gouldson and Murphy, 1996; Huber, 1985):

- The restructuring of production and consumption towards ecological goals. This involves the development and diffusion of clean production technologies and decoupling economic development from the relevant resource inputs, resource use and emissions;
- 'Economising ecology' by placing an economic value on nature and introducing structural tax reform;
- Integrating environmental policy goals into other policy areas.

The potential for doing this is illustrated by the fact that other rationalities, such as social and labour struggles, have, in the past, imposed limits on a purely economic rationale for production and consumption. However, the proponents of ecological modernisation do not wish to assert the primacy of ecological over economic rationality, merely to assert the necessity (as with sustainable development) of giving the former equal weight.

In Huber's view, ecological modernisation, as the only way out of ecological crisis, involves *more* industrialisation, albeit with changed production and consumption. As such he overemphasises the industrial and technological aspects and neglects the social context within which these occur. The 'ecological switchover' is thus seen as a logical, necessary and inevitable stage in the development of the industrial system. Technological developments occur largely autonomously and act to determine change in industrial systems and their relations with the social and natural environment. The propulsive force of technological change means that the state has little role in redirecting the processes of production and consumption. Indeed, while the theory of ecological modernisation proposes that institutions can be restructured on ecological lines and away from a purely economic rationale, the theory is largely silent on the extent to which such institutions can be reformed – this remains open to empirical investigation (Mol, 1994). Hajer (1993) argues that there are two interpretations of ecological modernisation. First, a 'techno-corporatist' interpretation which emphasises the 'economisation of nature' and elitist decision-making structures and a second interpretation, closer to 'strong' sustainability, which not only stresses changes to production and consumption, but does so through greater democratisation, redistribution and social justice. Christoff (1996) has characterised the two ends of this spectrum as 'weak' and 'strong' versions of ecological modernisation (Table 1.2). Hajer (1995) develops this idea of 'strong' ecological modernisation as reflexive ecological modernisation, whereby political and economic development proceed on the basis of critical self-awareness involving public scrutiny and democratic control, while 'weak' ecological modernisation involves a lifeline for capitalist economies threatened by ecological crisis.

Table 1.2 Characteristics of 'weak' and 'strong' ecological modernisation

'Weak' ecological modernisation	'Strong' ecological modernisation
Technological solutions to environmental problems	Broad changes to institutional and economic structure of society incorporating ecological concerns
Technocratic/corporatist styles of policy making by scientific, economic and political elites	Open, democratic decision making with participation and involvement
Restricted to developed nations who use ecological modernisation to consolidate their global economic advantages	Concerned with the international dimensions of the environment and development
Imposes a single, closed-end framework on political and economic development	A more open-ended approach with no single view, but multiple possibilities with ecological modernisation providing orientation

Source: Adapted from Christoff (1996).

As a political agenda, ecological modernisation has three linked programmes:

- Compensation for environmental damage and the use of additional technologies to minimise the effects of growing production and consumption on the environment;
- A focus upon altering the processes of production and consumption, for example through the use of clean technologies and economic valuation;
- The dismantling and deindustrialisation of economies and a transformation towards small-scale units and a closer link between production and consumption.

Mol and Spaargaren (1993) cite the example of Dutch environmental policies which have shifted from the first approach to the second. This has involved moves to: close substance cycles and the chain from raw materials through to the production process and waste and recycling; conserving energy and improving the efficiency and utilisation of renewable energy sources; and improving the quality of production processes and resultant products. One basic tenet of ecological modernisation is that it will be supported by business as it involves financial advantage – it responds to environmental issues through notions of profitable enterprise (Harvey, 1996; Weale, 1992). This comes about through five forms. First, through reduced pollution and waste production resulting in greater business efficiency. Second, through avoiding future financial liabilities, such as the potential future cost of clearing up contaminated land. Third, creating a better environment has benefits for, and attractions to, a company's workforce. Fourth, through the sale of 'environmentally-friendly' products and services and fifth, through the sale of pollution prevention and abatement technologies (Dryzek, 1997). While this may raise the possibility of the transformation of capitalist economies, ecological modernisation is also liable, as a discourse, to be 'corrupted into yet another discursive representation of dominant forms of economic power' (Harvey, 1996: 382) resulting in greater dominance of global resources by transnational industry, national governments and 'big science' in the name of sustainability.

To date, ecological modernisation has concentrated on the potential for environmental reform at the 'meso-level of national governments, environmental movements, enterprises and labour organisations' (Mol and Spaargaren, 1993: 454). While sustainable development envisages the devolution of power up to the international scale and down to the local, ecological modernisation does not necessarily require de-emphasising the nation-state. From the perspective of this book, ecological modernisation has not been utilised to address the problems of local areas. One of the few references to the subnational scale in the literature is made by Spaargaren and Mol (1992), in a criticism of the works of 'counterproductivity theorists' (such as Commoner, Illich, Gorz and Bahro) for their emphasis on the need for greater local autonomy and severing links with world market and political relations as a prerequisite of a response to ecological crisis. Instead (following Giddens, 1990) Spaargaren and Mol propose that such insulae cannot exist within a globalised world economy which interlinks and networks different social contexts and localities. Thus, while they recognise that localisation may be desirable, 'the intensification of international social relations and the increasing level of time-space distancing within modern societies make the realisation of these goals in the context of local experiments, which are thought to be exempt from power relations and market forces operating on a worldwide basis, less plausible and realistic' (Spaargaren and Mol, 1992: 331).

Given the largely national focus, the role ascribed to the local level and the local state is circumscribed. In some accounts (Huber, 1985) the state should play no role in the switch to ecological modernisation as it will only hinder the development and diffusion of clean technologies. Mol and Spaargaren (1993), who argue that it seems difficult to imagine an ecological switchover without state intervention at various levels, have criticised this overly technical view. This may not involve a role for a strong bureaucratic state in ecological modernisation (although Harvey (1996) appears to equate ecological modernisation with the politics of corporate and state managerialism). 'Rather the role of the state in environmental policy [will have to] change from curative and reactive to preventive, from "closed" policy-making to participative policy-making, from centralised to decentralised and from dirigistic to contextually "steering" (Mol, 1994: 17). Ecological modernisation will require political commitment to a longer-term, more holistic approach to economic development and the environment, as well as involving non-state actors and social movements (Mol and Sonnenfeld, 2000). Ecological modernisation could therefore usefully be deployed to inform theory relating to the local scale. Thus Jänicke (1992) argues that the capacity of individual nation-states to undertake the 'ecological switch' is determined by a combination of their economic performance, innovative capacity, strategic capacity and consensual capacity. Jänicke (1992, 1997) uses the concept of environmental capacity to determine the conditions that encourage nation-states to address environmental problems and to determine the conditions that can result in both successful policy forms and implementation. It has been argued that those nation-states which conform most closely to the ideas of (albeit weak) ecological modernisation (the Netherlands, Germany, Norway, Sweden, Japan) are those which have consensual forms of government (Dryzek, 1997, although see Andersen 1994, 1997 for a critique of this view). The same capacities will exist at the subnational scale and a key area for investigation may be the strategic capacity that exists within local areas where:

> Strategic capacity means that environmental protection really becomes a 'cross-section function' of the administrative authorities. Generally speaking, strategic capacity must include the capacity to integrate partial sectors of the state with a view to new objectives, and to dismantle contradictions and deduce conflicts about objectives.
> (Jänicke, 1992: 84)

However, this theory of strategic capacity has been formulated at a national level – it fails to consider developments either at supranational or subnational levels. Some guidance as to how this could be developed to deal with subnational levels comes through Jänicke's (1997) later work which builds upon the theory of strategic capacity to produce a model of policy explanation. Capacity for action on environmental issues for Jänicke (1997: 9): 'defines the necessary structural conditions for successful environmental policy as well as the upper limit beyond which policy failure sets in even in the case of skillful, highly motivated and situatively well-placed proponents.' In this instance capacity defines both opportunities and barriers (Murphy, 1999). In this model, solutions to environmental problems are developed within structural framework conditions and within situative contexts, involving actors and strategies, together with institutional, economic and informational factors (Murphy, 1999). Strategies are the general approach taken to a problem – using environmental policies to address problems and achieve goals, for example. Actors (including individuals, pressure groups, third parties, etc.) are the opponents and proponents of special issues. The latter have relatively stable general interests and core beliefs and their capacity for action depends largely upon their strengths and competencies. An important dimension to this model is that the ability of actors to develop strategies can be significantly influenced by the situative context, such as economic recession, public awareness or a major pollution incident. The other component of the model comprises the structural framework conditions that provide the backdrop to the situative context, actors and strategies. These form the broad conditions of environmental action and give rise to an opportunity structure for actors consisting of:

- The cognitive-informational framework – i.e. the conditions under which environmental knowledge is produced, distributed, interpreted and applied;
- The political-institutional framework comprising the institutional and legal structures and institutionalised rules and norms in a society;
- The economic-technological framework including economic performance, technology levels and sectoral composition.

The resultant capacity of actors to act (and the success of their actions) will be influenced by the interplay of these frameworks in any given situation. In utilising this model to understand the implementation of environmental policy at the local scale (and its integration with economic development activities), we begin to move away from the notion that implementing such policies and devolving policy to lower levels is a relatively problem-free process. Much will depend upon local situative contexts and structural frameworks, as well as upon the composition of local actors and form of local strategies. While ecological modernisation is therefore useful in helping us to think through some of the changes that need to be made to current economies, it is thus only beginning to address the social processes involved through notions of situative contexts and local actors. In order to address these issues it is useful to draw upon another body of theory – that of urban regime theory which has become a central part of understanding the processes of local economic development and urban regeneration in recent years in political economy approaches.

Urban regime theory

It is clear from the ecological modernisation approach that local environmental policy-making processes need to be placed within a broader context of the governance and regulation of local economies. The problem of neglecting such a perspective is that in its absence there is a tendency to treat the 'environment' as a relatively self-contained and closed system, the constituent elements of which can be monitored, modelled and, subsequently, regulated with little regard to its interactions with 'external' economic and political systems. Both the sustainable development and ecological modernisation literatures would suggest that local level environmental policy needs an integrated and holistic approach to the environment and economic development (Gibbs, 1996, 1997). However, policies which attempt to achieve this have frequently failed, largely due to their failure to take into account the constituent properties of local economic and political systems. The fact that local environmental conflicts continue to occur around problems of externality (e.g. NIMBYism) or a failure to internalise transaction costs associated with environmental regulation suggests that the relationship between governance and environmental sustainability at the local level is far more complex than imagined within the sustainable development or ecological modernisation literatures.

It may be that, in terms of analysing the policy-making process, particular problems occur when those relationships are projected onto a spatial context. Traditionally, the relationship between environmental policy and space has been viewed in one of two ways: as an externality or as a cost. First, the spatial context can be incorporated as an 'externality effect'. Thus one rationale for decentralising environmental policy would assume that environmental impacts either already are, or in the future can be, confined within well-defined local or regional physical systems, such as ecosystems, watersheds and catchment areas. The trick in terms of policy then is to determine the boundaries of these systems and attempt to introduce policies and regulations consistent with those boundaries. Such policies and regulations would aim to internalise externalities arising from, for example, property and access rights, overlapping jurisdictional authority, and so forth. The failure to internalise externalities may cause conflicts that, in turn, highlight the problem of co-ordinating the territorial division of labour. Solutions to these conflicts usually involve higher tiers of government or the organisation of new levels of jurisdictional authority (Cox and Mair, 1991). In this context, conflicts around the environment are not so much eliminated as displaced territorially.

The second way of incorporating space into environmental policy making is to treat it like any other 'factor of production' – as a cost, either to business or to the consumer. Thus any business owner, consumer or property owner who by virtue of their location is impacted by environmental policies or regulation must factor in 'the environment' as a potential cost (or benefit) depending on the nature of the policy or regulation. In this context, the effectiveness of local environmental policy-making can be gauged by the extent to which local stakeholders are incorporated into the policy making process and full account is taken of their actual, or potential, transaction costs (for example, using contingent valuation methods). Transaction cost approaches to environmental policy making suggest that policy consensus is more likely to occur when participants have built up trust based on the frequency of local contacts or have long-standing interests in the locality (Angel *et al.*, 1995). Conversely, a failure to negotiate and internalise transaction costs can be a measure of the level of local resistance to a particular set of environmental policies or regulations (see Feldman and Jonas, 2000).

Problems of incorporating space into an analysis of the environmental policy-making process highlight an issue generic to the governance of local economies in capitalism, that is, the problem of managing the (spatial) division of labour. This problem has been addressed by urban regime theory, which for the most part treats it as an organisational issue (Elkin, 1987). Accordingly, conflicts arising from the division of labour between state and market actors may be contained by governing coalitions made up of whichever interest groups are the dominant or hegemonic forces in the locality (Stone, 1989). Since over the course of time any given locality will have developed distinctive economic and political institutions, local political regimes are historically contingent (Ramsay, 1996). The nature of the local political regime can change, as can its policies: 'urban regime theory asks how and under what conditions do different types of governing coalitions emerge, consolidate, and become hegemonic or devolve and transform' (Lauria, 1997: 1–2).

Urban regime concepts could therefore usefully be deployed to analyse the characteristics of local governing coalitions and their policies. In the case of local environmental policy regimes, this would focus on interactions between firms, local politicians, environmental enforcement agencies (both local and national) and environmental pressure groups, as well as a whole host of activities that may not immediately be thought of as having 'environmental' consequences. It might also be possible to examine specific institutional developments within a locality resulting from attempts to internalise externalities and transaction costs relating to specific environmental policies and regulations. Conversely, a failure to internalise such costs and externalities locally would provide some indication of the 'boundaries' of the local environmental policy regime, and the extent to which non-local powers and capacities come to bear upon the local policy process.

However, urban regime theory may not be able to say much about the longer-term stability and coherence of local environmental policy regimes. Indeed, one of its major weaknesses is its failure to link the structure of, as well as changes in, the local political regime to wider systems of accumulation, modes of regulation and state policy shifts (Lauria, 1997). It is especially weak on the spatial context, notably the impact of changing spatial divisions of labour on the form of governing coalitions and territorial structures in the state. As Cox (1997) suggests, regime theory has failed to theorise in an adequate fashion fundamental interests in local economic development and the contingent conditions – policies, regulations, state structures, etc. – under which those interests are realised locally. Dryzek has argued that an intelligent approach to environmental issues demands:

> A dynamic, structural-level analysis of the liberal capitalist economy, where it might be headed and what realistically might be done to alter this trajectory to more ecologically benign ends. For a confident and globally organised liberal capitalism mostly insensitive to environmental concerns is the dominant political fact of our times. Without such an analysis, we are reduced to wishful thinking about how things might be different.
> (Dryzek 1997: 198)

To address these weaknesses the next section of this chapter outlines a regulationist perspective on local political regimes to allow for a more contextualised analysis of 'the strategic capacities rooted in local institutional structure and organisations' (Jessop, 1997: 63) and the role of the national and supranational state in regime formation (Ward, 1997).

Regulation theory

Regulation theory is an attempt to integrate the structural dynamics of capitalism with the institutional forms of society. These institutional forms are 'the concrete expression, the embodiment of the structure of society in a given historical period' (Moulaert and Swyngedouw, 1992: 40). Regulation theory analyses society and its institutions at three, interactive levels: the mode of production; the regime of accumulation; and the mode of regulation (Moulaert and Swyngedouw, 1992). It has been argued that the regulation approach should be seen as a general research programme within radical political economy or a broad frame of reference to analyse contemporary shifts in the economy and society more generally (Bakshi et al., 1995; Jessop, 1994, 1995). In all its variants it seeks to explore 'how economic and non-economic procedures can be articulated to produce a relatively stable, coherent and dynamic framework which can in turn secure the expanded reproduction of capitalism' (Jessop, 1990: 176). The regulation approach considers that crises may play a rejuvenating role for continued capitalist development, rather than leading to terminal crisis. The Parisian regulation school distinguishes four main stages in advanced capitalist development (see Table 1.3). After the period of extensive accumulation and competitive regulation that extended from the mid-nineteenth century until the First World War and an inter-war period of crisis, an intensive Fordist regime emerged. The fourth, and most recent,

Table 1.3 Phases of regulation and accumulation

	To 1914	1918–39	1945–73	1974 to present
Accumulation system	Extensive	Emerging intensive	Intensive (Fordist regime)	Emerging flexible?
				Protracted crisis?
Mode of social regulation	Competitive	Crisis of competitive mode	Monopolistic	Crisis of monopolistic mode
			(Fordist–Keynesian mode)	Emerging neo-competitive? Neo-conservative? Neo-corporatist?

Source: Tickell and Peck (1992).

phase began in the 1970s and is frequently seen as a period of transition from Fordism to a new regime of accumulation and regulation (Stoker, 1990).

The mode of regulation describes the collection of social relations (such as social institutions, behavioural norms and habits, as well as specifically governmental regulation) which act together in regulating the contradictions of capital (Painter, 1991; Peck and Tickell, 1992). A mode of regulation will stabilise for a while, but ultimately the extant mode of regulation for a particular regime of accumulation cannot resolve the contradictions and gives way to a period of crisis and transformation. In these periods of crisis, new structural forms emerge, some of which may form the basis of a new period of stability and hence a new mode of regulation, others may be short-lived experiments. Regulation theory does not predict the exact form of an emerging regime of accumulation, but is a conceptual framework for understanding processes of capitalist growth, crisis and reproduction. It focuses on relationships, mainly at the macroeconomic level, between the accumulation process and the ensemble of institutional forms and practices which together comprise the mode of social regulation (Peck and Miyamachi, 1994). These institutional forms and practices guide and stabilise the accumulation process and create a temporary resolution to the crisis tendencies which are seen to be endemic in the accumulation process. The mode of regulation is neither predetermined nor inevitable, as structural forms are the outcome of social struggles and conflict (Painter, 1991). The mode of regulation is the means of institutionalising these struggles between competing interests leading to the bounds that reproduce and legitimate the balance between production and consumption within a particular regime of accumulation (Marsden *et al.*, 1993). This places emphasis on the relationships between the main social actors in these processes, including the local and national state which act as mediators (Hudson, 1994). However, change in the social mode of regulation may be an outcome or a cause of economic change. The latter point is particularly relevant when considering the environment within regulation theory. It could be argued that the growing adoption of sustainable development and policies based on concepts of ecological modernisation, indicates an increased social concern with the relevant balance between production and consumption.

Within research from a regulation approach there has been a growing recognition that work to date has concentrated upon the economic at the expense of the social, cultural and political (Bakshi *et al.*, 1995). While an approach which links the economic with the social, cultural and political may appear to be appropriate to incorporate environmental issues, relatively few attempts have been made to link the explanatory power of regulation theory with emergent environmental concerns, the main exceptions being the work of Lipietz (1992a and b), Altvater (1993) and Drummond and Marsden (1995). Both Lipietz and Altvater examine environmental concerns within the context of a changing regime of accumulation. The logic of a capitalist regime of accumulation founded on intensive growth and mass production for mass consumption has been to both produce and stimulate consumption to the maximum (Lipietz, 1992b). Industrial mass production not only required mass purchasing power and hence a Fordist social system of labour and remuneration, it also demanded massive supplies of raw materials and energy from the global economy. The reification of social relations – where people relate to one another with money and commodities on the market – causes natural constraints on production and consumption to disappear from the consciousness of society. Nature only becomes relevant when it imposes additional costs or disrupts human life (Altvater, 1993). Indeed, under Fordism, the 'implacable logic' of the system is that it is better to set about repairing any damage (thus boosting consumption) than not to pollute in the first place. The alternative, of reducing pollution by adding it to production costs, aggravates the supply-side crisis, with the result that it is seen as an unaffordable luxury – the choice is between jobs and the environment. Such an impasse is not simply a question of having taken an unfortunate technical decision. 'The ecological crisis highlights the interconnection and interweaving of all the subsystems which the functionalist approach of productivism tried to isolate and put into separate boxes' (Lipietz, 1992b: 55). In a market-led solution the consequences of production and consumption are introduced into the system of values in a manner such that they are handled as expenditure on 'defensive costs of growth'. Rebuilding the environment may thus itself become a field of capital accumulation, whereby the expansionist drive turns to reconstruct the environment itself as an artifact (Altvater, 1993). However, even if, for example, industrial waste is converted into use values to satisfy human needs, then it can only be done through fresh expenditure of energy and materials and the repairs thus become part of the problem. The alternative is to organise the transformation of energy and materials from the outset in such a way that unavoidable entropy increase is kept as low as possible and to build into the functioning of the economic system a series of imperatives which prevent ecological damage (Altvater, 1993). Indeed, this is precisely the argument which is advanced in policies for sustainable development and ecological modernisation.

The logic of this is that any attempt to resolve the current environmental crisis solely through market mechanisms will be inadequate. The internalisation of environmental effects only represents a stopgap, which does not compensate for the way in which natural conditions are altered by the throughput of materials and energy in production and consumption. Economic instruments may assist in decision-making and provide some relief, but existing environmental degradation may remain and, more importantly, production processes and consumer habits remain unchanged (Altvater, 1993). Because an ecological critique of political economy hinges on an analysis of use-value, as materials and energy are transformed during the creation of use-values, the economic-ecological impasse cannot simply be treated as an ethical problem with the solution in changes of behaviour, such as using less energy or leaving the natural environment in no worse a condition for future generations. Neither ethical nor purely market-led rules are adequate. Such rules are insufficient without institutionalised rules of ecological behaviour. These imperatives must be institutionalised and equipped with sanctions, so that they become behavioural constraints for everyone. The ordering principles of this new approach must be to develop not only technologies that use less energy and materials, but also new forms of production and consumption (Leff, 1995; O'Connor, 1994). That is, there is a need to develop new modes of regulation and production.

Taking a regulationist perspective indicates that such new forms could be developed, given that the outcomes of economic restructuring are far from necessary, but open to debate and shaping. There are opportunities to shape future development on a more sustainable basis. Past historical compromises have 'resolved' capitalist crises, which at the time could have developed in various ways, such as the conflict between Fascism, Communism and Social Democracy in the 1930s (Lipietz, 1992a). The latest crisis is as yet unresolved and we are in the middle of a period of conflict where debate is over what form the new compromise should take. The neo-liberal 'solution' of the 1980s and early 1990s not only had a number of social and economic problems (the polarisation of society, the widening gap between workers and firms; the return of business cycles; the way that the free trade spirit of neo-liberalism is a source of international instability) but, importantly, it continued to foster a use of the natural environment which would ultimately undermine its own basis (Lipietz, 1992a; Peck and Tickell, 1994b).

From a regulation theory perspective, an important issue is the extent to which recent political, social, cultural and economic moves towards sustainable development and ecological modernisation can be interpreted as part of an emergent mode of regulation. It can be suggested that the current patchwork of international environmental agreements, growing public awareness of environmental issues, the rise of 'green consumerism', corporate environmentalism and the incorporation of sustainable development and ecological modernisation into local and national economic policy represent constituent elements of a new mode of social regulation (McGrew, 1993). However, at present whether they will (or can) cohere into a mode of regulation is very much open to future shaping through social struggle and conflict. It may be more appropriate at present to see environmentalism as one of a number of alternative strategies of regulation or new collective wills which have the potential to have a radical impact upon the conditions of existence of a regime of accumulation (Jessop, 1990). That the outcome may not necessarily be environmentally benign is indicated by the fact that certain elements of capital are already using the concept of sustainable development for the continuation of a particular set of social relations (Blowers, 1993; Harvey, 1993). Its adoption as a concept by multinational capital may be a reflection of this (O'Connor, 1994) and there are arguments that in some sectors sustainable development is being deployed and co-opted as a legitimation measure (Bridge and McManus, 2000). Indeed, it has been argued that central parts of the green movement are being co-opted into global capitalism and those that refuse are marginalised (Sklair, 1994). At best such an approach by capital would involve the adoption of the 'very weak' or 'weak' sustainability position on the spectrum of sustainable development in Table 1.1.

It is increasingly apparent that the scale and consequences of environmental problems are such that the modest reforms currently on offer through weak(er) forms of sustainability and ecological modernisation are inadequate for the task (O'Riordan, 1992). Such reforms may create a momentum for more radical measures. If sustainable development or ecological modernisation becomes the guiding principle of economic development, then radical changes in the form and nature of capitalism will have to be proposed. At the very least, sustainable development and ecological modernisation measures are incompatible with the type of neoliberal, free market policies that have gained ground in many developed countries in recent years (Jacobs, 1991; Peck and Tickell, 1994b). Co-ordination, co-operation, equity and democratic involvement are essential features of these measures. As Redclift and Woodgate (1994: 53) point out: 'the "environmental age" in which we live has, as a central concern, to consider whether our ways of exploiting nature are sustainable under any existing political and economic system. The challenge today is to embark on revolutionary changes in the way we organise ourselves to exploit nature.' At present sustainable development, ecological modernisation and state action ('real regulation') to achieve it can largely be seen as a continuation of past measures to legitimate certain levels of environmental impact. New definitions of the acceptable level of environmental impact, and at a level which will necessitate major change in the operation of capitalism, could result from the incorporation of sustainable development and ecological modernisation into policy. Exactly how this is to be achieved remains unclear (McManus, 1994). One area of broad agreement, however, is that current economic and social processes are unlikely to lead to automatic adjustment. 'It follows from this that if sustainability is to be progressed it will be because it has been purposively and objectively promoted through policies informed and empowered by a substantive theory of what sustainable development must be and how it can be brought about and maintained' (Drummond and Marsden, 1995: 53). Regulation theory offers a theoretical context which can help to inform how such adjustments can be made. The mode of social regulation has a variety of forms, from the 'real regulation' of laws and concrete structures through to more intangible elements, such as values and norms of behaviour. Peck and Tickell (1992) suggest that five levels of abstraction can be identified (see Box 1.1). While these levels have neither potential nor meaning in isolation, they do allow a consideration of the potential 'intervention points' for action (Drummond and Marsden, 1995).

The existence of these levels of abstraction indicates that policies to integrate the economy and the environment will need to be promoted at a number of intervention points and at a number of spatial scales. Intervention at the level of regulatory forms may be easier to initiate and comprehend, but such concrete forms of intervention must be underpinned by complementary social values and norms, i.e. the mode of social regulation as a whole (Drummond and Marsden, 1995). As Flynn and Marsden (1995: 1186) point out, 'regulation may be formalised through the enactment of legislation or established socially through sets of social practices, backed up by political and/or economic power'. The value of taking a regulation approach lies in identifying the need for this totality of approach rather than upon any individual form or scale. The use of regulation theory suggests that 'real regulation' by itself will not be enough to move towards sustainable development, but that this must be combined with changes in values and attitudes (Goodwin et al., 1995). Regulation theory proposes that the form of social regulation and the balance between production and consumption is open to struggle between competing interests. The outcome of this struggle is uncertain such that 'the macroeconomics of the future may be based on a downward spiral of

> *Box 1.1* Modes of social regulation
>
> - The mode of social regulation (MSR) represents the concept in its most abstract form, as a generalised theoretical structure abstracted from the concrete conditions experienced in individual nation-states (e.g. competitive regulation, monopoly regulation).
> - Within each MSR, a certain set of regulatory functions must be dispensed in order for the accumulation system to be stabilised and reproduced (e.g. the regulation of business systems, formation of consumption norms).
> - The regulatory system is a more concrete and geographically specific manifestation of the abstract MSR, typically (although not necessarily) articulated at the level of the nation-state (e.g. US Keynesianism, Pax Britannica).
> - Regulatory functions are dispersed through the operation of regulatory mechanisms, specific to each regulation system, which are historically and geographically distinctive responses to the regulatory requirements of the accumulation system (e.g. mobilisation of labour power, codification of financial regulation).
> - Regulatory forms represent those concrete institutional structures through which regulatory mechanisms are realised, although there need not be a straightforward one to one correspondence between mechanism and form (e.g. local states, legislative systems).
>
> *Source:* Peck and Tickell (1992)

social and ecological competition, leading to recurrent financial, business and environmental crises, or an ecologically sustainable and macro-economically stable model' (Leborgne and Lipietz, 1992: 347). Regulation theory has a role to play here, not through defining a single, post-Fordist development path, but in raising questions about the ecological sustainability of different development options (Peck and Tickell, 1994a).

Shifts in values and attitudes will themselves be highly differentiated, 'in some areas and situations a more sustainable future will be promoted and in others resisted' (Flynn and Marsden, 1995: 1189). The spectrum of sustainability outlined in Table 1.1 may be associated with a spectrum of policy and social changes which attempt to address the challenge of sustainability. This could range from a 'business as usual' scenario, where very weak sustainability is promoted through purely market measures, through to weak sustainability, where market measures are combined with some measures of international cooperation and agreement, through to stronger sustainability, where international agreements and binding protocols are produced, together with national and local implementation. Combined with major changes occurring in individual and collective behaviour, this would effectively mean the establishment of a new regime of accumulation and mode of social regulation. It seems unlikely that weaker sustainability or ecological modernisation measures will be sufficient to resolve the structural challenges to the Fordist regime of accumulation posed by the major environmental issues, but the exact form of stronger measures still require clarification (Pezzey, 1992).

This will require a reconstitution of the mode of regulation (and mode of production through ecological modernisation, for example, adopting clean technologies and recycling products) in order to achieve sustainable development. However, the regulatory mode is usually concerned to perpetuate the existing socio-economic order and to defend the distribution of power within that order. A fundamental barrier to an institutionalised solution is a doubt as to:

> Whether further accumulation is ecologically and economically sustainable. Based as it is on the transformation of nature, capitalism requires that nature is, effectively, an infinite resource. Yet as environmental resources are progressively degraded and as the end is in sight for oil (the commodity which both literally and metaphorically fueled Fordism), it is becoming increasingly clear that capitalism has perilously transformed all of nature.
> (Peck and Tickell, 1994a: 307)

Hence any strategy which seeks to redefine the object of regulation is necessarily radical because it provides a challenge to that social order. In consequence existing power structures form a major barrier to promoting the sustainable development and ecological modernisation agendas.

The mode of social regulation, as has been shown, has a number of levels of abstraction. The more concrete 'regulatory forms' are the easiest to understand and the most common locus of policy, for example at the level of the national or local state. However, it is unlikely that sustainable development can be promoted solely through intervention at one particular level. This is for two reasons. First, it is not possible to have a change at one level alone – change must come about *throughout* the mode of social regulation. Second, measures taken at any one level will be partial and have limited effectiveness and, indeed, may be counterproductive. In total then, different levels or 'sites of intervention' can not be considered in isolation. The emergent regulatory problems associated with whatever comes after Fordism will call for putative solutions based at different scales. Many of these will require regional, national and supranational co-operation. Local strategies, for example, may well have a role, but these must be located within a supportive national and supranational framework (Peck and Tickell, 1994a).

If environmental problems (such as global warming and sea level change) are left unchecked, the potential impacts upon global economies and societies are of such an order of magnitude that they will necessitate a whole new approach to concepts of economic development. While these impacts are increasingly recognised and attempts have been made at all spatial scales to introduce elements of a new regulatory system (although not in any co-ordinated fashion), it is clear that these attempts are themselves failing. In some cases attempts at environmental improvement have been overridden by other policy imperatives. This is especially notable in the area of trade policy, for example the ways in which world trade policy is environmentally detrimental or the impact of the Single European Market upon transport emissions (Lang and Hines, 1993). The benefits of drawing upon a regulationist approach are that regulation theory makes clear the non-deterministic nature of post-Fordism and that the future form of the economy is open to shaping and debate. It emphasises the need to consider both economic and social processes as an integrated whole and it indicates that sustainable development will need to be promoted at a variety of levels and scales (Table 1.4 sketches out some problems and some possible regulatory solutions). Thus 'capitalism may have inviolable laws but is has a plurality of logics, some of which may be more accordant with a sustainable mode of production than others' (Drummond and Marsden, 1995: 56). In practical terms, drawing upon such political economy approaches could contribute to devising appropriate policy outcomes. In theoretical terms there is a need to investigate the creation of the institutional basis of sustainable economies or the form of the mode of social regulation associated with sustainable development and to examine whether this can cohere to resolve the crisis of capitalism that stems from environmental problems.

Conclusions

This chapter began with a concern that the emphasis placed upon the local and regional scale in environmental policy-making has frequently rested upon rather vague notions of why these are the most appropriate scales for action. Following on from this it was posited that a lack of theoretical investigation into the rationale for local-scale action to integrate economic development and the environment has hindered the implementation of such policies. Both the literatures on sustainable development and ecological modernisation largely neglect directly confronting the question of the appropriate spatial scale for action. In the former approach, the local scale is simply assumed to be the most appropriate. In the latter approach the emphasis is upon the capacity of nation-states to act. The need for a more

Table 1.4 Environmental issues and possible regulatory solutions

Spatial scale	Environmental issues	Regulatory solutions
Global	Global warming Sea level change Free trade regimes Volatile monetary system	International pollution agreements Trade regulation reforms Monetary regulation Hegemonic power needed?
Supranational	Free markets Cross-border pollution Pollution emissions	Greater local integration Carbon taxes Clean technologies Different travel modes
National	Consumption patterns Pollution emissions Unemployment Short-term financial system	Recycling, reuse, repair Changing consumption attitudes Clean technologies More public transport use Restricted car use Carbon tax and reduced employment taxes Environmental job schemes Longer investment time scales New systems of national accounting
Local	Contaminated land Air pollution Lack of participation	Subsidy and reclamation Changing travel modes Local environmental indicators Environmental fora Greater democratisation Integrated economies

theoretical approach to environmental problems arises in part because it is increasingly evident that the implementation of policy is about the exercise of political and economic power (Owens, 1994). Indeed, while this is rarely made explicit by its proponents, sustainable development 'is a concept that is fundamentally *political*. Its realisation lies in answers to such questions as who is in control, who sets agendas, who allocates resources, who mediates disputes, who sets the rules of the game' (Wilbanks, 1994: 544). Sustainability is an ideological and political question, rather than simply an ecological and economic one (O'Connor, 1994). To my way of thinking, this highlights the relevance of concepts drawn from political economy for interpreting environmental policy, even though in neither case is there a substantive literature dealing with these issues. Thus the advantages of drawing upon concepts from regulation theory and urban regime approaches is that they allow some evaluation of the real possibilities for change at vulnerable locations in the political economy, rather than simply speculating on the potential for sweeping structural changes (Dryzek, 1994).

As currently constituted, modes of social regulation prioritise the value of capital and existing class structures while marginalising the material basis of sustainability. By defining the object of regulation in this way, society legitimates and empowers a set of causal mechanisms which sustain wealth and privilege at the expense of materially and morally unsustainable outcomes. This need not be the case – modes of social regulation are *socially* produced and reproduced and can thus be changed. But it is important to remember that they are constructed through a process of experimentation, struggle and conflict as opposed to being objectively promoted. However, while modes of social regulation can not be objectively constructed *per se*, the core values and institutions which legitimate and empower the mechanisms which underpin unsustainable outcomes can be changed. The problem though, which sustainable development and ecological modernisation approaches largely neglect, is that any strategy to redefine the object of regulation in this way is necessarily radical through its challenge to the existing social order. The fact that regulation is normally realised through existing power structures represents a major barrier to promoting such an agenda. But, sustainability can only be built around value and institutional shifts in society. This said, it cannot simply be the values placed on the environment which must change, but also the values and institutions which prioritise the value of capital and the maintenance of existing patterns of social relations (Drummond and Marsden, 1998). In Chapters 2 and 3 these more abstract concerns are examined in relation to the real instances of changing economic and environmental policy contexts.

2

THE CHANGING CONTEXT FOR LOCAL ECONOMIC DEVELOPMENT

Introduction

This chapter outlines the context for local economic development policy and initiatives. In Chapter 1, the perspectives examined emphasised the need to observe attempts to integrate economic development and the environment within broader policy and spatial contexts. To this end, in this chapter I examine some of the major shifts in the context for local economic policy making in recent years. One of the key themes here is the supposed greater globalisation of the world economy and the loss of sovereignty by nation-states. Capacity to act has purportedly passed upwards to international and supranational institutions and downwards to the local or regional scale – this process has been referred to as the 'hollowing out' of the nation-state (Jessop, 1994) or as 'glocalisation' (Swyngedouw, 1997). However, the powers of the nation-state have not vanished, but have been reconstituted and restructured as states seek to develop coherent strategies to deal with a globalising world (Giddens, 1990). Related to this have been changes in trade policies and the adoption of new technologies. In both cases, these are a response to, and a consequence of, globalisation. While providing the relevant context, my concern here, though, is less an examination of key trends in the global and national economies, than to provide an evaluation of the limits and opportunities these changes present in attempts to make local areas more sustainable.

As we have seen, there has been a growing interest in environmental issues at the level of international, national and local policy making. A major shift in thinking has embraced sustainable development and ecological modernisation as organising principles that allow a reconciliation between economic development and environmental protection. Much of this policy advocates taking local action to implement sustainable development, with a particular emphasis upon the role of local authorities as delivery agents. At the same time, however, there has been a major change in the form and structure of local economic development strategies gravitating towards greater emphases upon entrepreneurship, competitiveness, increased private sector involvement and a move from local *government* to local *governance*. More broadly, we can situate these changes at the local level within their international and national contexts. In both of these, there has been a move towards the greater integration of national economies into a global market, an increased emphasis on free trade and free markets, the growth of increasingly powerful global corporations and a political shift (until recently) towards neo-liberal governments and policies. In some cases these changes have taken on specific institutional forms, in particular the World Trade Organisation (WTO) and the proposed (but so far failed) Multilateral Agreement on Investment (MAI).

My argument here is that these broader changes to the economic context for local areas may well negate some of the policy options adopted at this scale or at least be incompatible with aims for greater sustainability. For example, policies for local sustainability not only require major changes in the ways that industry operates within an area, envisaging a shift towards 'clean production', but are also posited upon a base of greater equity and democratic involvement. By comparison, the shift to greater local competitiveness has produced greater inequality within and between areas, and changing local governance structures have reduced local democratic accountability. The consequences of this interplay between current policies for sustainable development and economic regeneration are likely to be continued environmental degradation and the profligate use of resources, rather than a local policy for sustainable development. Conversely, it can be argued that some of these changing contexts present policy makers with opportunities for following more sustainable development paths and for demonstrating the potential to introduce such policies at wider spatial scales.

Changing international and national contexts

A number of economic and political processes form the strategic context within which local actors currently pursue initiatives for sustainable development and these provide both opportunities and constraints. Those processes deemed here to be of greatest importance are: a shift from a Fordist to a so-called post-Fordist paradigm; the impact of new information and communication technologies; the globalisation of the world economy and related changes in trade policy; and changes in public policy away from welfare economics – what Jessop (1994) has termed a shift from the 'Keynesian Welfare State' to the 'Schumpeterian Workfare State' – and towards a greater emphasis on competitiveness and entrepreneurialism. Obviously these are interlinked in many ways, but they are examined here individually for the sake of clarity.

From Fordism to post-Fordism?

It has been argued at length that there has been a paradigmatic shift from a model of Fordist growth which was focused on mass production, scale economies and mass consumption to one focused on flexible production, economies of scope and more differentiated patterns of consumption. The implications of this shift at different spatial scales have also been much debated (Peck and Tickell, 1994b). In an overview of this literature, Bassett (1996) provides a useful summary of possible outcomes at the local level. These are:

- Optimistic post-Fordism: where the hollowing out of the nation-state and the breakdown of Fordism provide opportunities for local development. Such arguments are also predicated upon the observed decentralisation and localisation of production systems and the development of 'new industrial spaces' which supposedly create the opportunity for local areas to 'fix' mobile investment and construct neo-Marshallian industrial districts. Such neo-Schumpeterian arguments also stress the role of entrepreneurship and innovation (Porter, 1990) and the need for developing institutional capacity at the local level.
- Pessimistic post-Fordism: where the scope for local intervention is limited. This suggests, in contradiction to the more optimistic scenarios, that there is little evidence of the decentralisation of production systems or the emergence of stable neo-Marshallian districts. Instead local economies are increasingly dominated by transnational capital and most localities have little leverage over their investment patterns. In this scenario, post-Fordism leads to greater competition between areas, forced adaptation to global forces, reduced welfare and greater inequality.
- Finally, there is a scenario that envisages *a period of disorder and crisis* instead of any stable post-Fordist form, as implied in the other two scenarios. Localities struggle to exist in a post-Fordist regulatory crisis where economic development success is limited to the few and where fragmentation and disorganisation of both policy and politics is the order of the day.

A common approach by environmentalists has been to conflate some of the observed locational and organisational changes in the economy with the inevitable outcomes of a move to an optimistic post-Fordist scenario (see, for example, World Commission on Environment and Development, 1987; Friends of the Earth, 1990; Elkin *et al.*, 1991). Broader arguments for a restructuring of the world economy and 'to protect the local, globally' also rest upon the greater localisation of economic activity (Hines, 2000). It has been argued that the growth of small firms and smaller units of larger firms is an essential part of a post-Fordist restructuring of industry towards a form of flexible specialisation which will of necessity incorporate total quality management and environmental management strategies (Welford and Gouldson, 1993). Thus the decentralisation of industry provides the opportunity for a revival of small-scale, community and co-operative business which is thought to be somehow naturally more environmentally-friendly than large business. An example of this viewpoint is given by Crabtree:

> Local people who control their own local economy are less likely to waste their resources and pollute their environment than distant decision makers with no local roots – the integration of economic and ideological goals is best secured by decentralising the management of resources.
>
> (Crabtree, 1990: 177)

However, the environmental benefits of such assumptions largely rest on unsubstantiated assumptions. The basis for such conclusions are rarely stated implicitly and the main substance for such environmentalists' arguments is an implicit use of work by the flexible specialisation institutionalists (for example, Piore and Sabel, 1984; Hirst and Zeitlin, 1992) which has been the subject of a substantial critique (Amin and Robins, 1990; Amin, 1992; Amin and Malmberg, 1992). While it can be argued that the development of smaller scale and decentralised industry has been a major trend within this post-Fordist shift, this does not necessarily translate into independent small firms.[1] Large firms have been engaged in a process of disintegration and devolution to smaller units. Production may be decentralised, but control need not be. The decentralisation of control and local autonomy may therefore be limited. Moreover, decentralisation, where it has occurred, has been from older core industrial areas to 'new spaces' of industrial production. However, the development of new industrial spaces suggests that a new pattern of industrial agglomeration is occurring, rather than widespread dispersal.

Moreover, the local nature of production does not prevent problems arising such as the generation of local pollutants which contribute to global problems. The question of whether a smaller unit or firm leads to greater environmental responsiveness is a highly debatable one (Roberts, 1995). Ironically, this may be the case for small *units* of environmentally-aware corporations, but small firms do not necessarily have a good environmental record. In some cases their compliance rate with pollution restrictions and enforcing regulations, for example, are much poorer than large firms (Tilley and Fuller, 2000). Not only may large firms have the resources to invest in processes and products that have less impact on the environment, but they may be, for instance, more energy efficient or better able to utilise pollution abatement technologies due to their economies of scale (Welford and Gouldson, 1993). In some areas, such as waste disposal, small firms may be squeezed out of the market by increasingly stringent regulations. In the UK, small firm suppliers to larger firms may find that they are required to meet certain environmental management standards (such as ISO 14000) if they are to continue their supply function. This may improve their environmental performance, but the main driver of change is coming from their larger customers. Certainly any policy for environmental improvement or sustainable development must address the issues of large firms, not only because of their importance for the global economy, but also because they could be in the forefront of moves towards sustainability, rather than necessarily acting as a hindrance (Schmidheiny, 1992). Enthusiasm for small firms and local production is a normative view of the economy, and one that is frequently based on a rather romantic view of industrialism than upon fact.

An over-emphasis on some specific concrete forms has thus given rise to an argument that these are a necessary outcome of a specific type of post-Fordist economy and, moreover, one which ignores the alternative post-Fordist scenarios outlined at the beginning of this section. This argument has a number of problems. First, it assumes that these developments are inevitable and that incorporating sustainable development into policy aims will be going with the grain of economic restructuring, albeit remaining a difficult task. Second, it places a strong emphasis on changes in production as opposed to consumption issues. Third, it ignores the social processes underlying economic restructuring and the role of public policy in articulating these.

The impact of information and communication technologies

It has frequently been argued that advanced economies have entered an age where knowledge has become the basis of economic development and prosperity (Gillespie *et al.*, 1989; Hepworth, 1990). This knowledge takes the form of information, which has become a commodity to be exchanged, sold, transferred and used. As such, the application of information has transformed the economic base of advanced economies, so that some commentators have argued that we are entering the 'information economy' (Newton, 1992). While definitions of what constitutes the 'information economy' are subject to debate (see, for example, Webster, 1994), the term is used to refer to three particular aspects of structural change in advanced economies (Hepworth, 1990):

- The growing contribution of information-related activities to wealth generation and employment;
- The increasing centrality of new information technology, as a form of capital, in management, production and consumption processes;

- Higher levels of specialisation based upon the commodification of information, involving particularly the privatisation of public information and the externalisation of 'in-house' information services.

Increasingly, then, the basis of economic development has come to rest on the application of new information and communication technologies (ICTs). It has been suggested that these technologies will form the prime basis for future economic development, with approximately 60 per cent of all employment and 7 per cent of gross domestic product (GDP) within the European Union becoming either directly or indirectly dependent upon them (Ungerer, 1990). These technologies are both embedded in products and processes, but also act as the means to organise and control corporations and firms. The adoption of such technologies is closely linked to the development of flexible production and to the globalisation of the world economy, as interlinked computers and communication systems allow the organisation and co-ordination of production, marketing, distribution and management across space.

Some authors have sought to demonstrate that the trend towards greater ICT use will have positive impacts upon the environment and contribute to sustainability. Information and communication technologies are said to offer new ways of working and make working at, or closer to, home a real possibility with consequent reductions in pollution: 'advanced ... technologies will help to bring productive work back into the home and neighbourhood, and enable local work to meet a greater proportion of local needs than today' (Robertson, 1986: 87). For example, the Brundtland Report proposed that: 'new technologies in communications, information, and process control allow the establishment of small-scale, decentralised, widely dispersed industries, thus reducing levels of pollution and other impacts on the local environment' (World Commission on Environment and Development, 1987: 215).

This theme of the dispersal of industry is a recurrent one in much of the literature by environmentalists. In a reiteration of the optimistic post-Fordist scenario, dispersal is thought to involve local control (Galtung, 1986). Such local control is proposed to mean less resource use and pollution than if control operates through distant decision-makers with no local roots (Crabtree, 1990). A more self-reliant set of local economies will emerge that combine local control with reduced environmental impacts (Dauncey, 1986). Indeed, the whole notion of reduced travel in economic activity through technology-assisted dispersal is an important one. Porritt (1990), for example, has argued that the use of ICTs would mean an end to commuting, except for occasional meetings, as most long distance communication will be by face-to-face communication systems. Such ideas have been the subject of pilot projects to decrease long-distance commuting (see, for example, State of California, 1990). In certain areas of the US, such as California, Oregon and Washington State, there has been state level action to promote telecommuting. In Oregon, for example, the Employee Commute Operations (ECO) rules administered by the Oregon Department of Environmental Quality required employers in the Portland metropolitan area with more than fifty employees to develop strategies to reduce the number of single driver commuting journeys by their staff and imposed financial penalties for failure to implement such schemes (Bristow et al., 1997). The state also provided a 35 per cent business income tax credit on telecommuting equipment provided by companies which established telecommuting facilities for employees at their homes or in satellite offices. The US Department of Transportation envisages that eventually around 15 million workers will become telecommuters – around 10 per cent of the entire US workforce (Office of Technology Assessment, 1994).

In addition to these teleworking or telecommuting schemes, other proposed ways in which travel (and the resultant pollution) can be reduced include teleconferencing and, through reducing customer travel, in developments such as telebanking, call centres and telephone booking. For example, it has been estimated that promoting home delivery services could reduce car-based shopping trips by between 30–40 per cent (Friends of the Earth, 1997). Another theme is the notion that such developments allow the creation and/or retention of jobs and the provision of services in more remote areas, help to prevent out-migration, provide enhanced public sector services, such as health services, and encourage dispersed economic development. Again, the state of Oregon developed a programme of providing distance learning, teleworking and telemedicine facilities at Klamath Falls, a small town in south central Oregon, with the aim of reversing economic decline. It has also been argued that technology is changing the way in which consumption occurs, with a shift from the purchase of tangible objects to the dematerialisation of the economy with associated environmental benefits – for example, replacing CD purchase by downloadable music files from the internet (Leadbetter, 2000). This dematerialisation is also associated with the spread of corporate leasing behaviour replacing owning fixed assets, whereby producers retain responsibility for the product including disposal. It is proposed that 'this gives manufacturers an incentive to make it [the product] as robust and durable as possible, extending its life, and to recycle once its useful life is over' (Leadbetter and Willis, 2001: 38).

However, not all the supposed benefits from the greater use of ICTs have been demonstrated. It has been estimated that the impact of telecommuting in the UK would be minimal in terms of energy consumption – for every 1 per cent of the UK population who telecommute, there is only a national energy saving of 0.06 per cent (British Telecommunications, 1992). Additionally, there are a number of negative environmental impacts, including the growth of call centres on greenfield sites and the amount of car-based transport involved in staffing such centres for 24 hours a day, 7 days a week, the loss of vitality in traditional town centres with associated decay and decline, and the hypothesis that increased electronic contact actually leads to increased physical contact, often associated with non-work related travel, with an increase in travel and pollution (Graham and Marvin, 1996). Benefits from any dematerialisation of the economy may be outweighed by the growth of different forms of consumer goods. In addition, a major area of concern within research on ICTs is the notion of inequitable access to the new technologies, whether at spatial, institutional, organisational or social scales. This concern has also been raised by those investigating the relationship between ICT usage and sustainable development, given that sustainable development is based upon notions of equity and participation, as well as environmental factors (Bristow *et al.*, 1997). Finally, ICTs, through permitting the growth of just-in-time production systems,[2] act to increase the volume of road traffic with a concomitant rise in pollution.

Globalisation and the drive for free trade

The rhetoric of globalisation sees global economic processes becoming more powerful than nation-states, national economies subsumed into a global economy, and where distinct national management regimes of labour rights, social policy, fiscal and monetary policy are subordinate to international financial markets and transnational companies (Hirst and Thompson, 1996). The increasing growth of transnational corporations has seen their influence on terms of trade increase so that it has been argued: '"free trade" thus guarantees that national needs, values, conditions, environmental rules, regulations and standards – particularly those more protective of public health, fair labour and fair pricing standards – no longer apply to transnational business behaviour' (Long Island Progressive Coalition, 1997: 22). Such transnationals control more than 70 per cent of total international trade and have benefited markedly from the increase in the liberalisation of global trade and capital flows from the 1980s onwards (Welford, 1997). This growth of corporate power, it is argued, markedly decreases the accountability of private enterprise to workers, managers and local communities. Instead, localities have to engage with the process of globalisation on terms not of their own making. The key features are the need to be competitive, provide a low-wage, yet well-trained workforce, together with incentives to large corporations either to invest there, or to remain there and expand once the initial investment decision has been taken. In some cases this has not been enough to retain certain fractions of capital. In the United States, for example, some firms have moved to duty-free trade areas in Mexico, rather than remain in the higher waged, higher taxed and more heavily regulated United States (Churchill and Worthington, 1995). Even where companies remain, increased economic uncertainty and job insecurity act to discipline both workers and localities.

The free market economic models promoted by multilateral financial institutions and free trade associations have supported such global domination by transnationals. The North American Free Trade Agreement (NAFTA) and the Uruguay Round of the General Agreement on Tariffs and Trade (GATT), which established the World Trade Organisation (WTO), aim to reduce those barriers which impede corporate access to a country's workers and natural resources. Activist groups in particular have argued that these provide 'corporations with a closed door setting in which to argue the details of further de-regulation in front of un-elected judges in an unaccountable, non-democratic process to which the citizens of the world have no access' (Long Island Progressive Coalition, 1997: 28). The drive towards free trade is thus an integral part, if not the driving force behind globalisation. The process of international trade can have negative (and positive) impacts on the environment and related issues (see Box 2.1).

Box 2.1 Impacts of international trade on the environment

- Trade creates a dependency on the ecosystems of foreign countries as goods consumed in one place are either manufactured in another or where resources are derived from another.
- The consumption of goods imported from a country will also have an environmental impact in the country in which they are consumed (e.g. through the use of energy or the creation of a waste).
- Trade itself has a direct impact on the environment, mainly through impacts associated with transportation.
- Trade can transfer technology and employment from one location to another having a social and environmental impact on host and home country.
- Trade and economic globalisation have greatly expanded the opportunities for the rich to pass on their environmental burdens to the poor by exporting both waste and polluting factories.
- Trade has a secondary impact on issues of equity as it results in impacts (positive or negative) on poverty, health, employment, human rights, democracy, labour laws and self-determination.

Source: Welford (1997)

From a conventional economic growth perspective, trade allows growth and the transfer of technology, which subsequently allow environmental problems to be addressed.

The North American Free Trade Agreement, which went into effect on 1 January 1994, was the first trade pact to incorporate environmental provisions, though an environmental side agreement (formally known as the North American Agreement on Environmental Co-operation). This element of the pact was subject to heated political struggle, with severe opposition from business interests and a view amongst environmentalists that the provisions were too weak to be effective (Doughman and DiMento, 2001). Proponents of NAFTA within the US Congress based their argument upon the need for the US to embrace the irreversible logic of globalisation. While employment would be lost through trade-driven restructuring, this was seen as an inevitable process, with or without NAFTA. The solution was to go for economic liberalisation and the creation of a larger free trade market, which would boost trade, sales and, ultimately, employment (Churchill and Worthington, 1995). Essentially, NAFTA was predicated on the belief that GNP growth equates to improved human existence, that such growth based on free trade would eventually 'trickle down' to all members of society and that growth has no limits. Past evidence would suggest that for participants such as Mexico the real environmental consequences of past 'economic growth at all costs' have been water depletion and pollution, soil erosion and air pollution (Churchill and Worthington, 1995). Despite this, environmental quality initially received only minimal attention in NAFTA and then only through pressure from grassroots organisations. In the US, engagement with environmental concerns by both proponents and opponents of NAFTA in Congress largely focused on the environmental consequences within Mexico, rather than upon the potential impacts of increased growth in the US or Canada (Benton and Short, 1999). However, there were wider issues to consider where the fear was that:

> The mobilisation of US capital and technology with Canadian and Mexican resources in an integrated regional economy will deepen a polluting and unsustainable fossil fuel economy, while the enhanced freedom for USA-based corporations that monopolize finance and technology in many industries will exacerbate economic inequality.
> (Churchill and Worthington, 1995: 94)

In common with other trade and economic development policies, NAFTA promotes economic growth first, while the fruits of economic growth can deal with environmental concerns. As many authors have pointed out, this creates the anomaly that the clean up of pollution is actually recorded as beneficial for economies as it contributes to a growth in GDP (see, for example, Lang and Hines, 1993). Unrestrained consumption and increased energy use as a consequence of increased trade have not been addressed, while environmental action has been directed to the consequences of industrialisation (through 'end-of-pipe' measures, for example) rather than the causes. Such misgivings over the environmental impact of NAFTA have fed through into opposition for recent moves towards a Free Trade of the Americas Agreement (FTAA). Certainly the evidence would suggest at best a very mixed record of NAFTA's material impacts upon environmental quality in North America with loss of jobs, worsening pollution and adverse trends in the transport and hazardous waste sectors (Doughman and DiMento, 2001). In total then:

> The corporate agenda which normally dominates policy formation is readily visible in NAFTA, and members (of Congress) were largely subservient to it by formulating the problem as one of maximising growth to maintain employment in a fiercely competitive global economy. Environmental concerns were thus relegated to the subordinate space which they customarily occupy on the policy agenda, generating ineffective "side" agreements that were little more than a sop to pacify critics.
> (Churchill and Worthington, 1995: 97)

It has been argued that nation-states have the power to rectify this situation and take back some control over international trade and transnational corporations, if only they were committed to sustainable development. Casagrande and Welford argue:

> Individual states should have the right to introduce product norms and standards which regulate the local environmental effects of consumption and which ban products which have been produced in such a way which has an intolerable impact on the environment, human health or basic human rights.
> (Casagrande and Welford, 1997: 153)

Indeed, two of the critics of the 'rhetoric of globalisation' have been at pains to point out that even if the world economy is becoming, in their terms, more internationalised, the nation-state retains certain key governance functions (Hirst and Thompson, 1996). Moreover, they argue, there are governance possibilities in the form of international regulatory authorities and trade blocs, such as NAFTA or the EU, to enforce environmental objectives alongside economic and trade objectives.

The problem, however, is that there appears to be little sign of such governance solutions to the 'new world economic (dis)order' and this dislocation is particularly acute with regard to the environment. Indeed, existing international institutions such as the World Trade Organisation (WTO) established in 1995 compound the situation. The WTO has a mandate to eliminate barriers to the free movement of goods and services and effectively operates as a representative of the needs of transnationals against the intrusions of democratic governments (Korten, 1995). The WTO has the ability to insist that local and national standards do not exceed WTO-accepted international standards and allows for challenges if higher regulatory standards are imposed. In effect this means that if a country wishes to introduce higher environmental standards on an imported product than have been internationally agreed, it is not able to do so and must conform to WTO standards. Measures that restrict export of a country's own resources for conservation purposes, measures that require local processing of resources or give preference to local investors over foreign investors are also liable to be ruled as unfair trade practices. The interests of international trade thus take precedence over local or national laws (Korten, 1995). If subnational governments in signatory countries take such measures they they are also bound to comply with WTO rules, even though they are not direct signatories themselves. Countries that fail to comply face financial penalties and/or trade sanctions. Korten, gives an example of how this might be used by transnational corporations to their benefit:

> A US company growing fruit in Mexico uses a pesticide that leaves a toxic residue on the fruit that complies with the international standard but is greater than the standard of the state of California. The corporation might convince the Mexican government to bring a case under WTO. California would have no right to appeal an unfavourable WTO decision in either California or US courts.
> (Korten, 1995: 175)

Initial fears that the WTO would lead to such decisions were confirmed by a decision in 1998 by the WTO court of appeal in Geneva. This ruled that a US embargo of shrimps from India, Pakistan, Malaysia and Thailand, on the grounds that fishermen in these countries were killing large numbers of turtles through their failure to use excluder devices, was illegal. This was despite the fact that the US government argued that it was obliged to protect the turtles under the Bonn Convention on Migrating Species, the Biodiversity Convention and the Convention on the International Trade in Endangered Species (CITES). The WTO court ruled that these treaties and obligations to protect endangered species were overruled by the imperatives of free trade, although it did conclude that the US legislation was justifiable in principle under Article XX of GATT (ENDS, 1999).

Growing realisation of the environmental and social impacts consequent upon such trade policies has led to a growing grassroots opposition movement lobbying to overturn free trade provisions. Meetings of the WTO and the G8 (G7 plus Russia) group of countries have been beset by both street protest and some national government opposition in Seattle, Prague and Naples. To date the impact of this opposition has been fairly limited, albeit that it has led national governments to consider the implications of the international agreements they sign.[3] One area of success has been the stalled attempt to introduce the Multilateral Agreement on Investment (MAI). This looked set to further *restrict* the options available to nation-states and local areas, whether they adopt sustainability or not. The MAI was an agreement originally negotiated in the Organisation for Economic Co-operation and Development (OECD) which aimed to remove government regulations on foreign investment. As with other trade agreements, the rationale was that liberalising capital flows and removing restrictions on investment would bring benefits to all. Negotiations began in 1995 and were due to be completed in 1997, but this date was pushed back in the face of both internal disputes and growing external opposition and the eventual (possibly temporary) abandonment of the MAI in 1998. The proposed MAI had three key principles:

- Non-discrimination – foreign investors would be subject to the same treatment as domestic companies;
- No entry conditions – national and local governments could not restrict foreign investment in either sectors (except defence) or forms (e.g. privatised companies);
- No conditions – national and local governments could not impose 'performance requirements', so as to ensure local employment, control currency speculation or require a minimum period for investment. Such conditions were prohibited even if they applied to both domestic and foreign companies.

If national or local governments were in breach of these provisions, they could be taken to an international tribunal by transnationals and sued for past and potential damages. If nation-states had signed the MAI they would not have been able to withdraw for five years and would have been bound by its provisions for fifteen years.

While transnational companies would have gained from the MAI, there were fears that nations, regions and local authorities would have lost powers over local economic development and risk being sued under the MAI (World Development Movement, 1997). In environmental terms, transnationals would have been able to use the MAI to sue governments over any law or policy that impacted upon them more heavily than a nationally owned company. While there were proposals to include an environmental clause in the MAI, it was argued that the possibility of being sued would have dissuaded governments and local authorities from introducing regulations to protect the environment and local communities (World Development Movement, 1997). Again, the example of NAFTA is not encouraging, where an environment clause did not prevent the US-based Ethyl Corporation from suing the Canadian government for US$250 million over a Canadian ban on the toxin MMT (a petrol additive) for which the Ethyl Corporation is the sole manufacturer in Canada. At the regional and local level, it is possible that much local and regional legislation could have been ruled out under the MAI. In the US, this prompted the US government to push for the exemption of all US state and local government legislation. If not exempt, MAI provisions would have prevented local authorities and regional development agencies from using policies that, for example, ensured the employment of local people or use local suppliers, as well as undermining local level initiatives to promote sustainable development. While the MAI was eventually abandoned in its existing format, activists remain concerned that the provisions contained within it remain on the international policy agenda. In total, then, the shift towards a more global economy and associated trade liberalisation may well impose severe limits on what can be achieved in environmental terms at both national and subnational levels. There has been a growing concern at governmental level that some form of regulation should be introduced to control global financial trading due to the deleterious impacts of currency movements on nation-states (the so-called 'Tobin tax'). However, the pressure for similar restrictions on globalisation and trade for environmental purposes has predominantly come from non-governmental organizations and remains largely marginal to the international and national agenda, despite a G8 statement supporting the inclusion of environmental concerns in future WTO negotiations and the existence of the WTO Committee on Trade and the Environment (ENDS, 1999).

Changing public policies: neo-liberalism and the state

These trade policy-induced changes in the world economy have been concurrent with other, related, changes that have had profound implications for public policy within nation-states. Both national and local governments have greatly diminished their commitment to improving and providing welfare support. In the United States there have been cutbacks in the provision of public education, income support to low income families, publicly financed health care (Medicare and Medicaid), environmental protection and other social programmes (Pincetl, 1999). From the 1980s, this strand of neo-liberalism in politics was found in many advanced economic nations, taking various forms, popularly associated with particular individuals – 'Reaganomics' in the United States, 'Thatcherism' in the UK and 'Rogernomics' in New Zealand, for example. These national government strategies had in common the view that government should withdraw to a minimalist position and that business should be set free from regulation. Privatisation of public services, reduced welfare expenditures and a call for deregulation of health, occupational and environmental standards in order to be competitive in the 'unstoppable' global market were similarly common features. In the UK in particular, this meant that any interventionist strategies were ruled out in favour of a search for economic competitiveness based on a low-wage, flexible economy (Hay and Jessop, 1995). From the late 1990s, the election of less conservative, centre left governments in the US, the UK and across continental Europe toned down the harshness of neo-liberal rhetoric, but the mantras of globalisation and free trade remain unchallenged. For example, the task for the Blair and previous Clinton administrations, in the UK and US respectively, was defined as searching for a 'Third Way' which embraces free trade and globalisation, yet retains some semblance of social welfare. How far these two aims are compatible remains to be seen. In the USA the election of George W. Bush may result in declining social welfare provision and a much reduced environmental programme.

From an economic perspective, the dominant policy response from the 1980s onwards has been to argue that nation-states and local economies must become more competitive. Hirst and Thompson (1996) argue that this viewpoint means that the role of nation-states is to provide the infrastructure and public goods needed by business at the lowest possible cost. Within this broad statement, there has been an emphasis on attracting mobile investment, both through direct financial inducement and through marketing nations, cities and regions to potential investors. The adoption of new technologies, and the provision of an adequate telecommunications infrastructure, plus a higher skilled and better trained workforce, yet a low-cost and flexible one, are also part of the recipe adopted by most nation-states and localities. At the same time, there is increasing pressure to weaken environmental, health and safety regulations, lower job and income standards, reduce taxation and cutback social and welfare services in the drive to create and maintain a favourable climate for business. From an environmental perspective, however, it can be argued that such orthodox economic policies necessarily and inevitably create environmental degradation, the destruction of communities and the erosion of quality of life and living standards.

Although it is too simplistic to see changes in local governance and local economic policy as a straightforward consequence of globalisation processes, neo-liberal ideologies and the declining (or at least changing) power of the nation-state, there have certainly been related changes at the local scale. However, changing modes of regulation at the local scale are a constituent part of switches in the mode of regulation at the international and national scales. The extent to which local economies can resist the adverse impacts of globalisation has been open to much debate (Pacione, 1997). In an echo of environmentalists' arguments about potentially favourable post-Fordist scenarios, it has been proposed that the power of globalisation to transcend time and space could enable the development of a counter-hegemonic politics by facilitating decentralisation and democratisation (Agnew and Corbridge, 1995). Similarly Hall and Hubbard (1996) argue that cities are not the helpless pawns of international capital, but have the capacity to direct, if not control, their own destinies by exploiting their comparative advantages. The evidence for this is, as yet, slight, although it could be argued that recent resistance to trade agreements such as the WTO may indicate the possibilities for change, albeit that these have rarely been place-based to date. More obvious is the unequal relationship between the local and the global, such that we have powerful forces of global disorder on the one hand and 'largely reactive, and typically shallow, local responses' on the other hand (Peck and Tickell, 1994a: 298). Forms of resistance by local actors to global capital are undermined by the mobility of capital and the politics of place competition to which it gives rise (Cox and Mair, 1988). While some limited, local successes may be won, 'the general effect of globalisation is to reduce the power of local-regional states to promote progressive economic and social change' (Pacione, 1997: 1180).

The promotion of the local state through local initiatives for economic development has a history that stretches at least as far back as the late nineteenth and early twentieth centuries, through municipal welfare and enterprise schemes (Gough and Eisenschitz, 1996). Since the mid-1970s though, with the end of the long post-war boom, such local economic development strategies again gained prominence. Initially this reflected the failure of nation-states to deliver economic development, with rising unemployment occurring in many localities. In the UK, a number of local economic development initiatives were developed in the late 1970s and early 1980s by municipal governments controlled by the Labour Party – the best known of these being the London Industrial and Financial Strategies produced by the Greater London Council (1985, 1986). These initiatives were set up in opposition to the national Conservative government and had, as one objective at least, to indicate that there was an alternative to neo-liberalism, contrary to the views of Margaret Thatcher, Prime Minister at that time. As such, policies from the metropolitan counties in the UK[4] were always intended to act as exemplars for the type of policies that could be introduced by a progressive (at that time, read Labour) national government. With the continued ascendancy of the Conservative government in the 1980s and 1990s and changes to the form of central government support for local economic development, these challenges effectively withered away. Local level initiatives have continued though throughout the 1990s and 2000s as both Conservative administrations and the (New) Labour government encouraged a mix of partnership within, but competition between, local areas. Such 'new modes of governance' have been characterised by promoting local economic development, with the aims of preserving or increasing local prosperity, attracting inward investment and creating jobs (Hall and Hubbard, 1996). Rather than management of the city through bureaucratic means in the form of the local, there has been a move towards a broader mobilisation of local resources in the competition for development.

It is perhaps too simplistic to see developments since the 1980s as a totally new development in the history of local economic development and governance. Hall and Hubbard (1996: 155) point out, for example, 'that city governments, to a lesser or greater extent, have always pursued entrepreneurial strategies and played a crucial role in local economic development'. Similarly, Short (1999) outlines the important role played by 'boosterism' in the expansion of the frontier and early urban development in nineteenth-century America. Nonetheless, recent years have seen substantial changes in the context for, and form of, local economic policy making. In the UK, for example, central government policy in the 1980s and 1990s shifted away from regional policy towards urban aid and the creation of what has been termed 'central government localism' (Martin and Townroe, 1992), and a shift from welfare-based policies to place-based competitiveness (Stewart, 1994). In the case of the UK, therefore, it is difficult to see the emergence of entrepreneurial modes of governance in isolation from the broader changes in urban and regional policy under Conservative governments. 'The political rhetoric of the 1980s clearly positioned the private sector as the key actor in city rebuilding, with public-private partnerships presented as the way forward' (Hall and Hubbard, 1996: 157).

These changes have not been confined to the UK. The US, for example, has a longer history of such entrepreneurial local initiatives acting to combine the local state and business in partnership, in large part because of the fiscal weakness of the former. Regeneration in Detroit and Baltimore, for example, through the formation of 'urban growth coalitions' formed a key reference point for many localities in other countries. The US through the 1980s and 1990s displayed many of the features of the entrepreneurial mode of governance. National policies took a neo-liberal stance through facilitating capital mobility and cutting community programmes that might distort private investment decisions. The remaining community programmes were to be funded through 'performance partnerships', linking local funds to achieving benchmarks as opposed to local need (Clarke and Gaile, 1997). Increasing emphasis has been placed on the need to become competitive at the local level, not just *vis-à-vis* other areas within the nation-state, but internationally as well.

This shift in priorities is summed up by Lloyd and Meegan as being an approach where:

> It is the job of all who represent local interests, democratically or otherwise, to deliver supply-side advantage (cheap, flexible and skilled labour and desirable sites and infrastructure in a climate of entrepreneurship) to businesses and industries capable of holding their own in globalised competition.
> (Lloyd and Meegan, 1996: 75)

Conclusions

We should be wary of reading off such changes in local politics and governance from processes of macro-economic globalisation. It is useful to think of local decision makers selecting and interpreting certain elements of globalisation and directing their efforts to 'solving' these problems. Clarke and Gaile (1997) use the idea of 'causal stories' in relation to how local policy-makers view globalisation, its local impacts and the contested solutions to these problems. Hay and Jessop (1995) similarly argue that we need to map discursive shifts in the ways that the local economy and wider economic spaces are understood in order to understand the development of new modes of local governance and new regimes of local economic development. Such discourses or stories attach blame to certain aspects of the situation and privilege certain solutions over others. One interpretation sees globalisation as enabling the development of a 'new localism', as the ability to deliver the necessary local conditions of production required by global capital passes up to supranational bodies and down to the local scale – the notion of the 'hollowing out' of the nation-state or 'glocalisation' referred to in the introduction to this chapter. Liberalisation of trade similarly reduces the prospect of national intervention and flows of capital and commerce, creating 'new economic spaces' for local areas (Mayer, 1994). From such a perspective the task for localities is to find their niche in the global economic hierarchy. In this interpretation, the role of local areas, and particularly local governments, is to orchestrate development and upgrade their hierarchical position through strategic intervention. Under this scenario, the role of local authorities has shifted from being providers and guarantors of a level of welfare provision and supporting infrastructure for local electorates, to being enablers, intermediaries and change agents. Urban governance has been reoriented away from the local provision of welfare and services towards outward-oriented stances aimed at encouraging local growth and economic development (Hall and Hubbard, 1996).

Conversely there is a more problematic interpretation or 'causal story' around responses to globalisation, free trade and neo-liberalism which sees these changes as contributing to the erosion of both local authority powers and their level of control over the local economy. 'International trade liberalisation agreements such as the North American Free Trade Agreement reduce national authority, expose regions and cities to more global competition and constrain local autonomy' (Clarke and Gaile, 1997: 32) and allow the pre-emption of local powers in economic development and environmental regulation, amongst other powers. This form of the 'new localism' is characterised by the rise of local authority place-marketing strategies, the fragmentation of local governance and the emergence of private–public partnerships (Bovaird, 1994; Clarke and Stewart, 1994). Such developments are often categorised as indicating the rise of the entrepreneurial city, a form of local governance with a greater reliance upon private sector-led efforts to develop local entrepreneurial potential (Harvey, 1989). The more pessimistic view is that the inevitable outcome of this drive to create competitive advantage is that there will be winners and losers, creating a 'mosaic of unevenness' at every spatial scale (Lloyd and Meegan, 1996) and that local strategic interventions have little value in the face of global processes:

> It is difficult to see how local strategies can do anything other than bend to the will of global competition: progressive local social contracts are likely to be difficult to sustain in the face of 'jungle law' at the global level. The basic difficulty, then, lies in trying to establish local order in the face of global disorder.
> (Peck and Tickell, 1994a: 298)

The context for local economic policy making has therefore shifted to a scenario where localities need to adapt to broader processes, whether this is seen as an opportunity or as a threat. Greater emphasis is placed on the need to involve a broader group of actors in the policy-making process and the development of 'partnership approaches' to economic development. In the UK the response at local level has largely been to adopt more entrepreneurial strategies, using promotional campaigns and partnership approaches to encourage inward investment and economic development. In large part the homogeneity of the response is a product of the centralised nature of the UK state and the changing form of public policy on, and funding for, local regeneration (K. G. Ward, 1996, 1997). In essence the tightening of local government expenditure in the UK meant that there was little or no alternative to competing for private sector investment and entering into partnerships to draw down funding.[5] By contrast, the US shows a much more diverse pattern of local strategies which encompass more entrepreneurial strategies as in the UK, but which also attempt to counter the effects of globalisation rather than simply embracing it (see Table 2.1). In the US cases examined by Clarke and Gaile (1997), for

Table 2.1 Local strategies in the United States

Type of strategy	Main forms
Locational strategies	Factor costs important
	Policies to reduce factor costs
	Attracting inward investment
World class community orientations	Seeking a niche in the global economy
	Focus on innovation capacities and transaction costs
	Public–private partnership
	Information infrastructures
Entrepreneurial mercantile orientations	Diversified growth built on local initiative and indigenous assets
	Focus on new and small businesses
	Local self-reliance
	Community and non-profit institutions
Asset-based human capital strategies	Human capital strategies
	Creating jobs with public funds
	Linking forms with labour pools
Sustainable development	Focus on sustainability values
	Equity concerns
	Attacks on local pollution

Source: Adapted from Clarke and Gaile (1997).

example, the introduction of NAFTA had led the Cascadia alliance in the Pacific Northwest to develop a sustainable development strategy, in an attempt to reduce the environmental impacts of increased North–South trade links.

These developments in the policy framework for local economic development form the context for local initiatives intended to move towards *sustainable* local economic development. The outcome of these changes in the context for local economic development is the antithesis of those changes envisaged in policies for sustainable development. It has been argued that the shift to more entrepreneurial modes of governance has exacerbated social and spatial inequalities and neglected social justice in favour of enhancing the prosperity of certain elite groups (Hall and Hubbard, 1996). Here we can see a close relationship with the wider processes of globalisation. Both in global economic development and local economic development, the underlying ethos has changed to one where free market will deliver prosperity and wealth. As opposed to the growth of greater equity, democracy and participation, there has been a move towards greater inequality, privatisation and reduced democratic involvement. Castells (1997) highlights the conflict between these tendencies towards globalisation and the development of what he terms the 'space of flows', with environmentalists' focus on locality, self-sufficiency, participative democracy or the 'space of places'. This is a useful reminder of the history of environmental movements, with their roots in self-management ideals, small-scale production, self-sufficiency and a desire for 'the control over space, the assertion of place as a source of meaning' (Castells, 1997: 124).[6]

In the context of the search for sustainable development at the local level, the key feature of local economic development strategies has been the search for *growth* by whatever means. Hay and Jessop (1995) have argued that the predominant means of achieving this is predicated upon the favourable articulation of the local economy with the wider economy. How this plays out will vary from locale to locale, but perhaps the key feature to note here is that it is difficult for local strategies to develop around an environmental focus in the absence of a wider complementary economic framework. Similar conclusions have been reached in the context of developing new forms of national and regional governance (Amin and Thrift, 1995; Low et al., 2000). UK research has indicated that where local authorities have attempted to integrate economic development strategies with environmental issues, the latter frequently lose out to the former (Gibbs et al., 1996, 1998). In contrast to Gough and Eisenschitz (1996: 209) who argue that 'the growth of the "entrepreneurial local state" has helped to reduce traditional interdepartmental rivalries within the local authority', this would appear to be far from the case between economic development and environmental functions. Concern for environmental issues within local economic policy making, and especially sustainable development, is largely confined to environmental and planning departments within UK urban local authorities (Gibbs et al., 1996). In cases where environmental issues have been addressed in local economic development strategies, they have either been translated into weak forms of sustainability that either do not threaten the status quo, such as the clean-up of contaminated land, or are seen as directly complementary to entrepreneurial activities, such as the improvement of the physical environment to encourage inward investment. In these cases, the environment essentially becomes a commodity to be repackaged for corporate consumption (Katz, 1998). The entrepreneurial local area thus prioritises inward investment, property development and job creation over environmental targets, whilst the result of constructing new buildings and roads will be increased use of energy, transport and materials (Ravetz, 1996). In the case of some UK central government local regeneration initiatives, such as City Challenge and the Single Regeneration Budget (SRB), these attitudes have been given institutional force through viewing the environmental aims of economic development as targets for land reclamation, improvement and kilometres of roads built or improved. Indeed, the objectives embodied in initiatives such as SRB, which are very influential in shaping economic development schemes, act as a severe disincentive to incorporating sustainability objectives.

At present then, policies for sustainable development and economic development at the local scale appear to be incompatible. In the UK, for example, economic development policies have taken precedence over environmental concerns and the outcomes, in terms of built form and business formation, reflect this dominance. However, it is increasingly apparent that there are conflicts with economic development policies based around ideas of greater entrepreneurship and competitiveness. The emphasis on local competitiveness and economic growth, with a reliance on trickle down effects to bring the benefits of this growth to poorer sections of the community, has failed – frequently the costs rather than the benefits of growth are most likely to trickle down. Moreover, implementing such strategies will continue to lead to the degradation and exploitation of the environment. Changing modes of governance demand a re-engagement with questions of social justice and a consideration of whether entrepreneurial strategies produce a fair and defensible distribution of benefits and burdens in society.

One means of advancing this re-engagement is to revisit the need to integrate environmental issues and sustainable development within local economic development policy, rather than to see it as an extra mainly concerned with minor improvements to the physical environment. One suggestion has been that environmental functional regions should be devised, building on concepts of territorial integration, catchment areas and bioregions, akin to the notion of self-reliant cities (Cohen, 1995; Haughton, 1997). Similarly, Regional Planning Guidance within the UK could assist the implementation of central government's own sustainable development strategy (Council for the Protection of Rural England, 1994). To date it has largely failed to do so, due to:

> Poor opportunities for public participation; the exclusion of major urban areas (covered by strategic planning guidance) in some regions, resulting in 'doughnut' planning; the dominance of the roads programme; the failure to reflect regional distinctiveness; the failure to advance national environmental objectives; the centralisation of power in central government; and poor monitoring and review.
>
> (Glasson, 1995: 723)

Certainly there is a case for a much greater integration of environmental and economic issues at a variety of spatial scales. Local environmental 'solutions' are not sufficient, given the transboundary nature of many environmental problems. Creating a sustainable local economy should not rely on 'exporting unsustainability' elsewhere (Wackernagel and Rees, 1996; Roseland, 1997).

What is needed, then, is a reassessment of the interrelationship between international environmental polices and their implementation at national, regional and local scales, so that such polices are both implemented locally and yet informed by their operationalisation in a iterative process. As Roberts argues:

> It would be foolish to believe that all the actions that are necessary in order to bring about a change in business attitudes can be taken at a local level, or in isolation from the national and international dynamics of economic and ecological transformation. The internationalisation of economic activities, together with the widespread occurrence and consequences of environmental damage, implies the need for action to be taken at all spatial scales from the global to the local.
>
> (Roberts, 1995: 242)

Indeed, it can be argued that the key issue in governance, whether for environmental or other policies, is the need to ensure that the governing powers (international, national, local) are 'sutured' together in a relatively well-integrated system (Hirst and Thompson, 1996). Proponents of globalisation and free trade assume that the free market can play the role of providing adequate governance to ensure this occurs. More realistically, though, it will take a major effort of policy to ensure effective public governance. In the absence of this, it is unlikely that global market forces and transnational corporations can ensure integration. As we have seen, however, while the globalisation of the world economy may be an overstatement, in the sense of totally free markets subject to little or no regulation, the governance structures that do exist for global and supranational trade are either explicitly constructed to *exclude* environmental considerations or, at best, are likely to rule that free trade overrides any environmental concerns of local, regional and national governments.

At present environmental policies are devised and implemented in a piecemeal fashion, by a variety of agencies, and with little or no coordination. However, local level actions can only proceed within this broader framework. At the practical level, local policy initiatives need to be situated within an international and national policy framework and should not supersede regulatory action at the national or international scale. It is in such a context that policies for local sustainability can be rethought. Sustainable development is not just about incorporating environmental concerns into economic development decisions, but factors such as inter- and intra-generational equity and greater democratic involvement in decision making are also central concerns. Existing economic development policies, in the UK and elsewhere, may stress the development of local partnerships, but these are frequently interpreted simply as a partnership between the private sector, local authorities, and the newly created institutions of local governance. However, as Leonardi points out:

> Social capital cannot be established and accumulated if local communities are not empowered to determine their own outcomes. Programmes need to be developed to encourage self-reliance and collective action on the part of individual citizens and groups of citizens working together to achieve common goals.
> (Leonardi, 1995: 178)

Within a locality there is scope for basing a set of new and refocused economic development policies around sustainability. These could include targeting inward investment policies on environmental technology sectors, encouraging improved environmental standards through supply chains, offering demonstrations on waste management and pollution control to local firms, developing local exchange trading systems (LETS), creating jobs through environmental improvement schemes and devising local indicators of sustainability. Such developments should occur as an integrated holistic strategy, rather than as a set of isolated and disjointed initiatives. Both national and local policy makers need to take a more imaginative response to current problems, as opposed to reiterating the same solutions to local problems – a paralysing condition induced by notions of an entrepreneurial state and the mantra of competitiveness. After a consideration of the environmental policy context in Chapter 3, I turn to an analysis of some of the attempts that have been made to develop such integrated policy strategies in Chapters 4 and 5.

Notes

1 Even in instances where independent small firms are in the ascendancy, there is a long-standing debate over the extent to which they are truly independent. Several authors have argued that most small firms are closely tied to the fortunes of their suppliers and/or customers. What matters is less legal independent status than the power relationships between firms (Taylor *et al.*, 1995).
2 Just-in-time production systems are closely associated with the development of flexible production systems and a post-Fordist economy. Basically, supplies of components, sub-assemblies, parts, etc. are delivered to the factory as they are needed, rather than the producer holding large stocks of these on-site. This system allows greater flexibility in what is produced, as well as reducing the costs of holding such stocks through reduced warehousing and the high cost of large inventory.

3 That legislators may not always have read or understood the implications of the laws they approve was illustrated vividly by Ralph Nader's offer of $10,000 to the favourite charity of any member of the US Congress who had read the 800-page document approving WTO provisions and could correctly answer some simple questions in public. Only one member came forward to do so and as a consequence subsequently voted against implementing the provisions (Goldsmith, 2001).
4 The metropolitan counties in the UK were abolished in 1986 in large part because of Conservative opposition to their development of oppositional forms of 'municipal socialism'.
5 The most striking example of this in the UK context was the example of Manchester City Council which moved from a hard-line 'municipal socialism' in the early 1980s, in opposition to central government, to an entrepreneurial partnership approach by the end of the decade without any substantive change in elected local officials. The change was exemplified in the changing slogans used in council literature and on council property – from 'Defending Jobs, Improving Services' to 'Manchester – Making It Happen' (Peck and Tickell, 1994a; Bassett, 1996).
6 Castells (1997) usefully qualifies this by observing that environmental movements may be 'local' in terms of space, but they are frequently 'global' in their approach to time, echoing both notions of futurity in concepts of sustainable development and the old environmental slogan 'think global, act local'.

3

THE CHANGING ENVIRONMENTAL POLICY CONTEXT FOR LOCAL ACTION

Introduction

This chapter examines the environmental policy context for attempts to integrate economic development and the environment at the local level. As with the economic policy context outlined in Chapter 2, this creates both opportunities for, and barriers to, local level actions. In certain cases it may provide some element of legitimation for local areas in nation-states which have limited national environmental policy initiatives. This was a particular feature in the UK in the 1980s and early 1990s, where local authorities used both international and European Union (EU) environmental policy to legitimate their activities in the face of a hostile, or at best indifferent, central government. The 1992 United Nations Earth Summit was a key event in this context. As Blowers (1994: 169) indicates 'the broader international context ... increasingly sets the parameters for policy-making at subsidiary national and regional levels'. The Earth Summit provided the momentum for local actors in some countries to develop their own environmental agendas by reference back to this international agreement. This chapter deals specifically with environmental policies, as opposed to providing a review of *all* policies that have an environmental impact. Attempting such an overview is beyond the scope of this book, as it would necessitate a detailed study of international and national policy across a broad range of areas, such as fiscal, social and development policies. Rather, the aim of the chapter is to provide an obverse view to Chapter 2 through examining environmental policy with a view to its consequences for economic development at the local scale. I begin with an overview of the main international agreements on the environment in recent years, then proceed to an examination of policy within the European Union. EU environmental policy perhaps represents the most detailed attempt to shift patterns of economic development onto a more environmentally aware basis at the supranational level and, both because of and despite its flaws, is worth examining in some detail. Finally, I outline varying policy contexts within such differing nation-states as the

USA, Japan, Australia, Sweden and the United Kingdom, both as an examination of national policies in their own right, but also to place the scope for local level action in context.

The international policy context

International agreements on the environment

As Selman (1996) indicates, the interconnectedness of environmental systems requires a concerted international response. By comparison with the international agreements on trade examined in Chapter 2, however, international environmental government and governance are poorly developed, despite a number of high profile agreements intended to influence national and local strategies. Such global initiatives can be divided into 'multi-sectoral' programmes, dealing with the environment in an holistic fashion, and 'sectoral' or 'thematic' strategies, which address single issues (Selman, 1996, after Carew-Reid et al., 1994). In either case, such initiatives tend not to have specific legal and funding measures, but may be instrumental in generating controls and incentives at national levels.

Although, as indicated in Chapter 1, sustainable development is a concept that has been around in various guises for several years, it acquired popular momentum with the publication in 1987 of *Our Common Future*, the report of the United Nations World Commission on Environment and Development (WCED), or the Brundtland Report as it is often called. Sustainable development, in the much quoted definition utilised by the WCED, 'meets the needs of the present, without compromising the ability of future generations to meet their own needs' (WCED, 1987: 43). This commitment to sustainable development at an international level was reiterated at the United Nations Commission on Environment and Development's major conference held in Rio de Janeiro in 1992 – the Earth Summit – resulting in the UN's Agenda 21 programme. This set an agenda across a wide range of policy and management issues, based on the need to integrate economic, environmental and social issues. The Earth Summit led to three global statements on environmental concerns:

- The Framework Convention on Climate Change;
- The Convention on Biological Diversity;
- The statement on Principles of Forest Management.

The Earth Summit led to the establishment of the UN Commission on Sustainable Development and the Earth Council to monitor follow-up and compliance with the conference's action plan – Agenda 21. In addition to outlining the contribution of various groups in society (women, young people, business, for example) Agenda 21 set out a specific series of objectives to be achieved by local authorities at the international, national, regional and local scales:

- By 1993, a consultative process aimed at increasing dialogue between local authorities at the international scale should have been initiated;
- By 1994 co-operation and co-ordination between associations of cities and local authorities should have increased and facilitated the exchange of information and experience;
- By 1996, most local authorities should have undertaken a consultative process with their populations and achieved a consensus on a Local Agenda 21 for their communities.

The Agenda 21 documents from the Earth Summit stressed the role of local actors, particularly local authorities, in delivering the sustainable development strategy. For example:

- Two-thirds of actions from the Agenda 21 documents required the active involvement of local authorities;
- Local government is one of Agenda 21's nine named 'main groups', with a specific chapter (Chapter 28) devoted to its role;
- Local authorities are called on to produce Local Agenda 21 strategies as a key part of the Agenda 21 process.

Local government was seen as the level of government closest to people, and thus had a key role in educating and mobilising the public around sustainable development. This was to be achieved through an extensive programme of consultation with local stakeholders to inform a local programme of capacity building to achieve sustainable development – the Local Agenda 21 (LA21) process (Keating, 1993). It can be argued that Agenda 21 marked a major success for local authorities in being formally recognised as a key part of the sustainability debate (Mehra, 1997).

The explicit need to integrate economic development strategies with environmental concerns was recognised in Agenda 21. Thus:

> Sustainable development requires a commitment to sound economic policies and management, an effective and predictable public administration, the integration of environmental concerns into decision-making and progress towards democratic government, in the light of country-specific conditions, which allows for full participation of all parties concerned.
> (United Nations, 1993: 49)

This has been exemplified in calls for an 'eco-industrial revolution' to redefine the goals of industrialisation and the means of achieving it. This process of industrial restructuring for sustainable development has two main priorities (Robins and Trisoglio, 1992):

- Redirecting corporate activity to satisfy human aspects of development by refocusing product and process development to meet basic human needs, open up decision making within business to the wider community and creating sustainable livelihoods;
- Maximising long run efficiency in the use of environmental resources in the production and consumption of goods and services, through a move towards 'closed' industrial ecosystems, favouring renewable or recyclable resources.

Effectively then, the Earth Summit proposed an ecological modernisation approach to reconciling economic development and environmental protection. How this was to be achieved remained rather more vague, although the literature produced as a consequence of the Earth Summit emphasises that this shift will only be achieved through a process of co-operation and partnership between industry, government and a wider set of stakeholders than has previously been the case. Thus:

> governments need to provide an empowering framework, and within this stimulate, cajole and sometimes force a long-term shift in industrial behaviour. This will involve the application of a range of policy tools at different levels from the global to the local.
> (Robins and Trisoglio, 1992: 159)

This reflects the inherent weakness of Agenda 21 in that despite being signed by 178 national governments, it has no legal sanctions, no financial guarantees and no strong means of implementation (Elander and Lidskog, 2000).

Much less well publicised than the Brundtland Report and the Rio Earth Summit was the UN's Habitat II conference, or the 'City Summit', held in Istanbul in 1996. This conference focused upon the environmental, social and economic issues posed by urban development and produced a 'Global Plan of Action'. In economic terms this aimed to encourage the development of sustainable human settlements and developing economies that will make efficient use of resources within the carrying capacity of ecosystems, and by providing all people with equal opportunities for a healthy, safe and productive life in harmony with nature and cultural heritage and spiritual and cultural values, ensuring social progress. 'Fundamental to Habitat II is the notion that the future of the earth will be heavily determined by the quality of life in cities, i.e. urban policies become crucial' (Elander and Lidskog, 2000: 41). Again, local actors, and especially local government, were seen as a key focus for developing an 'enabling strategy' to ensure that such policy developments took place.

In 1997 a special session of the UN General Assembly reviewed progress with the Earth Summit's proposals. While this established a programme for continuing with Agenda 21, it failed to produce any political commitment and had little real impact, reflecting the loss of momentum post-Rio (Elander and Lidskog, 2000). Whether the Rio Plus Ten summit in 2002 gives rise to any new impetus remains to be seen. International meetings in recent years have focused more upon the specifics of climate change policy and greenhouse gas emission reduction (in Kyoto, Buenos Aires and The Hague) rather than upon the all-encompassing themes addressed in Rio and Istanbul.

The response to these international agreements has varied greatly. In some countries local authorities have been the main players in developing Local Agenda 21 plans and there has been an enthusiastic response (O'Riordan and Voisey, 1997). In other countries there has been only a limited participation in the LA21 process. The greatest contribution of these international policy agreements is probably their symbolic value. They indicate a level of commitment to the environment and sustainable development on the part of international and national policy makers, even if the rhetoric from the various conferences and agreements is greater than concrete results and actions. At the local and regional scale, their importance lies in the fact that they legitimate activities that may have no basis in statutory powers. Local and regional governments and other groups of local actors can call upon these agreements to justify their own actions, as well as using them as a stick to beat recalcitrant national governments.

The European environmental policy context

The integration of environmental issues into other policy areas has been a feature of the EU agenda for some time. As with other policy initiatives the actual success of this as opposed to the optimistic rhetoric contained within policy documentation remains elusive. However, attempts to shift the basis of development within the European Union towards *sustainable* development are of interest because it represents a rare attempt to develop and implement policy at the supranational scale. For member states, the influence of EU environmental policy upon *national* policy and its incorporation into other EU policy initiatives and funding streams is therefore of paramount importance. EU environmental policy is thus an important source of constraints on business and other activities (Hanf, 1996). The EU also acts as a means of translating international environmental policy agendas down to member state level – the EU is a signatory to the Earth Summit Agenda 21 protocol and to the Kyoto targets for emission reductions.

The principle of incorporating the goals of environmental policy into other policy spheres was formalised in the Maastricht Treaty in 1992 (see Box 3.1). The Treaty for the first time gave equal weight to the economy

> *Box 3.1* Changing approaches to the environment in the Maastricht Treaty
>
> - Promoting sustainable and non-inflationary growth, respecting the environment becomes one of the EU's guiding principles;
> - EU environmental policy to be based on the precautionary principle and preventative measures;
> - Environmental measures must be integrated into other EU policies;
> - Introducing qualified majority voting for some environmental measures speeds up agreement procedures;
> - Greater influence for the European Parliament in environmental legislation;
> - Tighter enforcement of environmental legislation through the Court of Justice.
>
> *Source:* Adapted from Department of the Environment (1992)

and environmental protection, Article 2 stating that the Community shall act to promote sustainable and non-inflationary growth respecting the environment. Article 130b directs environmental policy to contribute to the strengthening of economic and social cohesion. Specific environmental provisions within the Treaty on European Union are set out in Title XVI, Article 130r, which states that:

- Community policy on the environment shall aim at a high level of protection, taking into account the diversity of situations in the various regions of the Community;
- The precautionary principle will form the basis of policy, preventative action should be taken, environmental damage should be rectified at source, and the polluter should pay;
- Environmental protection requirements must be integrated into the definition and implementation of other policies.

The Commission's subsequent White Paper 'Growth, Competitiveness and Employment' (Commission of the European Communities, 1993) reflected this shift in emphasis and encouraged the adoption of sustainable development as a new model of development. Chapter 10 of the white paper saw economic growth being achieved in tandem with measures to increase employment and improve quality of life, alongside lower energy and natural resource consumption. The approach advocated was effectively one of ecological modernisation, with emphasis placed on increased efficiency, longer-life products, recycling and clean technologies in order to decouple the poor economic-environment relationship (Clement and Bachtler, 2000).

After much political wrangling over its functions and location, the EU also established the European Environment Agency (EEA) with an initial budget of 12 million euros and 25 staff. The EEA has the objective of providing EU institutions and member states with information to enable them to take measures to protect the environment, assess the results of such measures and ensure that the public is properly informed about the state of the environment and the pressures on it (Box 3.2). Finally, the Treaty of Amsterdam which came into force in 1999 established sustainable development as an explicit EU objective, making it clear that sustainability needed to be integrated into all EU policy areas.

Specific EU policy and action on the environment has occurred through Environmental Action Programmes (EAPs), the first beginning in 1973 and the most recently completed – the Fifth EAP – running from 1993 to 2000. While early programmes were essentially remedial in nature, focusing mainly on toxic waste disposal and the protection of public health, the Fifth EAP, entitled 'Towards Sustainability', stressed the need for a more proactive EU policy to alter behavioural patterns and encouraged a move towards sustainable development (Commission of the European Communities, 1992). The programme emphasised encouragement and promotion as

---

Box 3.2 Key tasks for the European Environment Agency

- Establish and co-ordinate an Environmental Information and Observation Network (EIONET);
- Provide objective information to enable the development of sound and effective environmental policies;
- Collect and analyse data on the state of the environment and provide uniform assessment criteria to determine compliance with legislation;
- Ensure environmental data are comparable;
- Disseminate reliable environmental information and publish reports on the state of Europe's environment;
- Promote the application of environmental forecasting techniques;
- Stimulate methods to assess the costs of environmental damage, protection and restoration;
- Stimulate information exchange on best available technologies.

Source: Adapted from *ENDS Report*, 240 (1995)

policy instruments, but also continued with the centralised regulatory approach adopted since the early 1970s. The programme was prepared in two main parts. The first provided a State of the Environment Report for Europe with an overview of environmental and natural resources in the EU, the pressures upon these resources and the trends in environmental quality and degradation. The second part of the programme was more concerned with setting objectives, together with policy and implementation programmes for the period to the year 2000. The EU argued that local actors were of central importance in implementing the Fifth EAP – for example, the European Commission estimated that implementing approximately 40 per cent of the Fifth EAP was the responsibility of local authorities (Hams and Morphet, 1994). The method of involving local actors was envisaged as being through a framework of co-operative partnership between local, national and EU levels (Hanf, 1996). A number of key areas were identified within the Fifth EAP where local authorities and other local actors could play a role (see Box 3.3).

The Fifth EAP differed from previous programmes in that it:

- Focused on those processes which deplete resources or damage the environment, rather than dealing with the consequences;
- Attempted to achieve a shift in behaviour patterns in society, especially in relation to consumption;
- Broadened the range of instruments available to achieve these policy aims – involving legislation, market-based instruments, 'horizontal' instruments (i.e. data and research and development) and financial support (e.g. LIFE, Structural Funds).[1]

Box 3.3 Key areas in the EU's Fifth Environmental Action Programme for local authorities

- Spatial planning
- Economic development
- Infrastructure development
- Control of industrial pollution
- Waste management
- Transport
- Public information, education and training
- Internal auditing

Source: Commission of the European Communities (1992)

Five target sectors were selected for particular attention under the EAP. These were chosen as areas where EU involvement was deemed appropriate or productive and where the environmental impacts are greatest: industry, energy, transport, agriculture and tourism. With regard to industry the EU envisaged a move away from an antagonistic approach towards dialogue and co-operation. Three issues were seen as particularly important:

- improved resource management;
- using information to promote consumer choice and confidence in industry;
- establishing EU standards for products and processes.

The European Commission's progress report on the Fifth Environmental Action Programme in 1996 maintained the focus on the five target sectors and singled out manufacturing as having made most progress towards sustainable development (Commission of the European Communities, 1996). Other sectors showed less progress – transport in particular continued to be problematic with rising traffic volumes and associated problems of congestion and vehicle emissions. Moreover, attempts to take more radical action to deal with these issues, such as the introduction of an EU-wide carbon tax, have so far been rejected by the Council of Ministers (Clement and Bachtler, 2000). In total then, while the European Commission argues that some progress has been made, the issues of climate change, pressure on natural resources, health and quality of life remain on the policy agenda. Indeed, one review of the Fifth EAP branded it as a failure (*ENDS Daily*, 8 January 2001). These shortcomings are being addressed through the EU's Sixth Environment Action Programme covering the period 2000–10, which is implicitly adopting an approach based on ecological modernisation principles:

> What we have to achieve is a de-coupling of the negative impacts on the environment and the consumption of natural resources from economic growth. De-coupling means economic growth while keeping the environment intact by a more efficient use of resources and higher environmental standards. By enhancing the eco-efficiency of our patterns of production and consumption, we will reduce the footprint of our society on this Planet.
> (European Commission, 1999: 22)

At a meeting of EU environment ministers in 1999 key policy measures proposed for the Sixth EAP included sectoral targets for improving eco-efficiency and new economic instruments to reduce resource consumption.

Environmental Action Programmes are also intended to encourage the integration of environmental issues into other policy areas. In addition to its own measures, the Fifth EAP was intended to influence the criteria for *all* EU funding and the programme therefore had important practical implications for local actors within Europe. These included particularly the impact upon the Structural Funds, i.e. the European Regional Development Fund (ERDF), the European Social Fund (ESF), the guidance element of the European Agricultural Guidance and Guarantee Fund (EAGGF-G) and, since 1993, the Financial Instrument of Fisheries Guidance (FIFG). In total, 146.2 billion euros were allocated for structural action for the 1994–99 period, with 9 per cent allocated to thirteen community initiatives such as LEADER, REGIS, RECHAR, KONVER, RETEX and URBAN[2] (European Commission, 1994b).

Some funds within the ERDF have already been directed towards environmental projects. For example, the Community Support Frameworks (CSFs) which form the contracts between member states and the Commission required conformity with environmental legislation and an assessment of the environmental impact of projects. However, the Environmental Impact Assessment (EIA) Directive (85/337) only applied to major development projects, and even here implementation has been uneven across the EU (ENDS, 1994). The requirement for EIAs has itself been criticised and arguments made that Strategic Environmental Assessments are needed in order to move away from a project-based approach. The revised Structural Fund Regulations adopted in 1993 require member states seeking Structural Fund support to produce environmental profiles of the regions in their regional development plans (Roberts and Jackson, 1999). This profile should contain a state of the environment report, an evaluation of the environmental impact of the plan, details of the arrangements for the participation of environmental authorities in the preparation of the plan, implementation and monitoring, and details of arrangements to ensure compliance with EU environmental legislation (Expert Group on the Urban Environment, 1996a). For eligible regions, the Regulation stated:

> Development plans for Objectives 1, 2 and 5b[3] must include an appraisal of the environmental situation of the region concerned and an evaluation of the environmental impact of the strategy and operations planned, in accordance with the principles of sustainable development and in agreement with the provisions of Community law in force.
> (Regulation 2081/93, 20 July 1993, OJ L193, 31 July 1993)

The EU has thus moved towards greater integration of environmental issues into funding decisions within the Structural Funds. The environment is increasingly being treated as a mainstream concern of the development process. For example:

- The ERDF encouraged a more integrated approach, taking full account of the environmental impacts at an early stage in formulating plans and programmes and avoiding environmental degradation.
- The ESF had increased support for job creation and training related to environmental activities that facilitate sustainable regional development.
- The EAGGF-G increasingly only provided support conditional upon the exploitation of agricultural and other rural resources in an environmentally acceptable way and eventually in a way which enhanced and improved the rural environment.

LIFE (the EU Financial Instrument for the Environment) provided opportunities for pilot projects from local actors. It has had a relatively modest budget though – about 90 million euros for the Community as a whole in 1996. It has been suggested by some Members of the European Parliament that LIFE should be extended into a proper Structural Fund. LIFE was intended to promote and demonstrate projects consistent with sustainable development, for example, through developing ways of measuring environmental quality and providing education and training. LIFE focused on projects involving co-operation between member states and actors, in line with a partnership approach (see Box 3.4 for an example of a LIFE project).

Despite these good intentions, environmental considerations have often been marginalised in regional development programmes, where job creation is deemed more important and where environmental protection often took the form of land reclamation and landscaping (Clement and Bachtler, 2000). Towards the end of the 1980s therefore the European Commission attempted to take a more positive approach towards environmental policy, not just in relation to the more traditional themes of creating attractive environments for inward investors, but also stressing the competitive advantage to be gained from developing environmental technologies, products and services and the job opportunities that this could involve.

While this has been a slow process, there has been an increased awareness of environmental issues within specific Directorates-General (DGs).[4] For example, the DG for employment has argued that environmental protection is an important motivating force for employment restructuring, whereby environmental programmes not only achieve ecological aims, but also strengthen the long-term competitiveness of regions and create jobs (Commission of the European Communities, 1990). In terms of regional policy, the Regional Policy Directorate of the Commission has stated:

> Box 3.4 The ISIS Project: an example of LIFE funding
>
> This LIFE project was intended to develop and clarify the concept of sustainability and apply it to real transport and local land use planning situations and choices. The Integrated System for Implementing Sustainability (ISIS) used a geographical information system (GIS) to construct a pollution and sustainability audit model which can help to compare the environmental effects of different transport and land-use planning scenarios. The partnership and collaborative approach is shown in the involvement of four partners reflecting a range of environmental and economic conditions: Kirklees (UK), Berlin (Germany), Copenhagen (Denmark) and Madeira (Portugal).
>
> In addition to the model, the project aimed to produce:
>
> - an integrated audit system on a GIS;
> - a computer-based visual demonstration package using GIS techniques to encourage participation;
> - good procedures manuals in sub-topic areas such as air pollution, noise pollution and social impacts of traffic;
> - an integrated audit manual, including measurement protocols, indicators and targets.
>
> *Source:* Adapted from Expert Group on the Urban Environment (1994)

A more integrated Community over the next decades must increasingly set its economic objectives in terms of sustainable growth. . . . Without a reversal of the environmental degradation of the past, the older industrial regions will be bypassed by regions encouraging modern, cleaner forms of activity. Similarly, in the regions lagging behind, environmentally insensitive development will damage their economic prospects at a time when many firms are actively seeking out environmentally attractive locations. For the stronger regions, the failure of the Community to achieve improved regional balance implies further congestion with all its associated environmental costs.

(Commission of the European Communities, 1991: 34)

The report *Europe 2000+ Co-operation for European Territorial Development*, which provided the reference framework for regional and local development over the 1994–99 period, stated:

> Safeguarding the environment and its biodiversity, as well as the fight against pollution should in future be a priority for spatial planning organisation ... and should inspire Union programmes and polices through systematic environmental impact studies, including at the planning stage.
>
> (European Commission, 1994a: 17)

The report went on to suggest that improving the environment would help to improve European competitiveness and foster inward investment. However, the report was essentially disappointing in terms of its commitment to, and understanding of, sustainable development in a regional and local economic development context. Most of the environmental emphasis in the report was directed solely towards more traditional environmental concerns, such as the protection of open spaces, water reserves and traffic congestion, rather than towards measures which link together the environment and economic development, such as job creation or industrial sector initiatives.

The example of the *Europe 2000+* report indicates that the integration of the environment into all aspects of European policy is far from complete. Indeed, despite the rise in importance of the environment within EU policy making, there remain a number of problematic areas:

- The implementation of this policy is, as yet, fairly weak. Environment policy has been enacted through Directives, leaving member states to decide how objectives should be achieved, thus creating a gap between those devising policy and those implementing it. The Commission itself has a poor record of reporting on the application of legislation (Institute for European Environmental Policy, 1994).
- Action Programmes do not have a budget – they are policy instruments used by the European Commission to encourage its own departments, authorities in member states and other partners to give more consideration to environmental issues.
- The uncertain future development of the EU calls into question exactly how environmental policy will be implemented. For example, there are major differences in attitudes towards environmental policy and the existing state of the environment in newer member states, such as Austria, Finland and Sweden, and potential member states such as Poland, Hungary and the Czech Republic (see Hallstrom, 1999).
- There is inconsistency of EU policies, even within Directorates. For example, the Regional Policy DG opposes 'environmentally insensitive development', but has continued to encourage and fund large-scale infrastructural projects (such as dams, new highways and wasteful irrigation schemes), especially in the Objective 1 regions. Sustainable development is still seen as a predominantly environmental concern, rather than an economic one.

- There are inconsistencies between establishing the Single European Market, encouraging greater global competitiveness and environmental policy. As with international trade policy examined in Chapter 2, Article 100a of the Maastricht Treaty specifically states that environmental solutions adopted by member states should not threaten the development of the Single Market. One example of the discrepancies this involves is the way in which free trade within the Single Market has encouraged greater volumes of road transport with a resultant increase in vehicle emissions.
- The Interim Review of the Fifth Environmental Action Programme published at the end of 1994 (Commission of the European Communities, 1994) concluded that progress has been slow in broadening the range of policy instruments away from the purely legislative approach and that the increased integration of environmental considerations into other policy areas, such as regional development, had also been too slow. This criticism extended into final assessments of the Fifth EAP, which deemed it to have been a failure.
- While integration with other policies was a key aim of the Fifth EAP, the Environment Directorate has had to rely on 'persuasion rather than power' with limited results (Wilkinson, 1997: 153).

Despite these problems, for local authorities and other agencies and actors involved in local development across Europe, taking account of both specific European environmental legislation and the way that it is being incorporated into other policy areas, such as regional development, is vital. Not only do such changes provide the context within which local policy and initiatives should develop, but also in practical terms an appreciation of the way in which these policy developments are affecting EU funding criteria is invaluable. This is particularly important given the power shift in the way that the Structural Funds have moved from being controlled largely by member states to more control by the Commission, to the preparation of regional development strategies within regions and the production of Single Programming Documents (SPDs) which identify regional problems and devise coherent regional programmes to address them, including an appraisal of the environmental situation. For all local actors, the European policy context means that whatever policies and programmes are adopted at the local level, they need to have sustainable development issues integrated within them. The EU's Interim Review of the Fifth EAP further emphasised the necessity of reconciling economic development with environmental protection and it is unlikely that these pressures will abate in future EU policies, despite the contradictions and failures of EU policies (Commission of the European Communities, 1994).

Given that EU policy will continue to stress the need to integrate these two policy areas, then simply seeing environmental issues or sustainable development as an 'add-on' or something that can be dealt with in isolation will not be sufficient. Instead, the incorporation of the environment needs to be an integral part of economic development strategies from the outset. In total, for local areas and regions, obtaining finance from the EU will increasingly require the incorporation of sustainable development issues into plans and programmes if they are to receive funding. Increasing control over the funds at the EU level, coupled with the expressed desire to form partnerships with the local level, shifts the emphasis to the local level to come up with ways of using finance that fit in with the 'greening' of the Structural Funds which was associated with amendments to EU regional policy following the publication of Agenda 2000 (Roberts and Jackson, 1999). All actions proposed in Structural Funds regional development plans must now be in accordance with the principles of sustainable development.

For those local and regional areas within the EU's constituent member states European policy looks set to increasingly influence actions to integrate economic development and the environment. Indeed, despite failures and problems with EU policy it represents one of the few attempts to direct economic development onto a more sustainable path. There is a whole set of important issues for the EU to deal with as it apparently moves towards greater economic and political integration, ranging from a single European currency to harmonisation of fiscal regimes to dealing with high levels of unemployment. One key issue is the extent to which Europe can move away from the type of free market, neo-liberal economic regimes which were dominant in the 1980s and early 1990s and engender a shift towards what is frequently referred to as a more 'Social Europe' and a new mode of regulation. Part of this is likely to involve much greater integration of environmental issues into economic development policy. In the context of differing integrative approaches to the environment and economy developed in Chapter 1, EU policies are closer to 'weak' forms of sustainability or the political programme of ecological modernisation, which see integration as an opportunity to develop new technologies, new forms of economic activity and create jobs (Gouldson and Murphy, 1996). Indeed, proposals for the EU's Sixth EAP argue that 'high environmental standards are ... an engine for innovation – creating new markets and business opportunities' (Commission of the European Communities, 2001: 9). Whether such policies *can* become part of mainstream policy or the dominant discourse in EU development remains to be seen.

Differing national contexts

EU policy is perhaps an exceptional case where supranational policy making is having a substantive impact upon local and national policy. Even here though, the influence of national governments, particularly at the implementation phase of policy, is of paramount importance (Gouldson and Murphy, 1996). In more general terms, while international agreements and policy are important, the national context of legislation, policy and government structures provides the main scope for local action (Selman, 1996). Indeed, the UN Earth Summit conference identified nation-states as the starting-point for the implementation of the Agenda 21 programme and proposed that: 'each country should aim to complete . . . if possible by 1994, a review of capacity, and capacity-building requirements, for devising national sustainable development strategies, including those for generating and implementing its own Agenda 21 programme' (United Nations Conference on Environment and Development, 1992: 200). Such national documents were seen as constituting national sustainable development strategies. In addition, signatories to the agreements on climate change, forestry and biodiversity were also expected to produce national policy statements on these issues.

Subsequent international meetings and resultant agreements have predominantly focused upon the need to reduce greenhouse gas emissions. In Kyoto in 1997, developed countries agreed to cut their emissions of such gases by an average of 5.2 per cent over the next fifteen years, while in Buenos Aires in 1998 discussion centred around the potential for further reductions and the means of achieving these. Given that these national contexts remain important, in the following section I examine the national and local environmental policy context in North America (USA), Pacific Asia (Australia and Japan) and Europe (UK and Sweden) to provide an indication of the range of possible responses and the ways in which this influences local attempts to integrate economic development and the environment. These countries were chosen to provide contrasting evidence of their engagement with environmental issues, ranging from high (Sweden) to low (the USA) levels of concern, as well as indicating changing attitudes over time consequent upon the changing economic context for national development. Given space constraints this section of the chapter is particularly concerned with policy developments during the late 1980s and 1990s in response to the sustainable development agenda, rather than attempting a comprehensive historical review of environmental policy in each country.

The USA

Environmental legislation in the United States has a substantial history of confrontation between environmentalists and industry. The Clean Air Act (1970), the establishment of the Environmental Protection Agency (EPA), the Water Pollution Control Act (1972) and the Surface Mining Control and Reclamation Act (1977) all institutionalised a confrontational stance between regulators and industry (Whitfield and Hart, 2000).

Environmental controls were thus seen by industry as a burden and this view was shared by the Reagan administration elected in 1980. Similarly, the subsequent presidency of George Bush was antipathetic to environmentalism, most notably in opposition to the climate change and biodiversity treaties at the 1992 Earth Summit. This stance softened with the election of the Clinton administration from 1992 onwards and the US showed some signs of evolving the more pragmatic and co-operative relationship between regulators and industry evident in other countries and the European Union (Wallace, 1995). The EPA introduced several programmes to reduce pollution and promote efficient resource use. These included plans to reward those companies which voluntarily reduced pollution and improved efficiency, a 'Green Lights' programme to encourage the installation of energy-efficient lighting and an initiative to lower computer-related electricity use (Corson, 1994). In addition, the National Brownfields Program 'attempts to promote sustainable development through regulatory and policy reforms that are designed to accelerate the redevelopment of idle, potentially contaminated commercial and industrial properties' (Tzoumis *et al.*, 1998). Overall, however, the US legislative system has been criticised for its impact on industry and economic development. For example, Wallace ascribes this to:

> The negative effects of a fragmented market, represented by the patchwork of regulators, unpredictability of regulatory requirements, caused by the slow gestation of primary legislation, about-faces in Congress and the courts and political interference with the EPA, and a single-medium focus which is biased against the all-round benefits typical of cleaner production technologies.
> (Wallace, 1995: 124)

A generally unreceptive climate for environmental legislation is evidenced by the weakness of Local Agenda 21 within the USA. Lake (2000: 71) concludes that 'Local Agenda 21 programs in the USA are few in number and widely scattered in non-central locations'. Lake, using data from the International Council for Local Environmental Initiatives (ICLEI) identified only 22 US local authorities engaged in Local Agenda 21 programmes, representing only 2.3 per cent of the total US population. Even this level of activity was confined to specific geographical areas of the US with virtually no LA21 programmes in the mid-west suggesting 'that the population loss and economic decline characterising this region are not conducive to discussions of sustainability' (Lake, 2000: 74). Any formal national programme along the lines of Local Agenda 21 has been confined to some of the activities of the President's Council on Sustainable Development (PCSD), established as an early initiative of the Clinton administration. The PCSD was intended to make explicit and popularise the link between regulatory reform of US environmental legislation and sustainable development. The PCSD operated through issue-specific task groups in an attempt to involve industry in the process of regulatory reform. For example, the climate change task force approved a set of Climate Principles which focused on three areas of policy development: the role of communities in climate mitigation; development and deployment of climate-friendly technologies; and incentives for early action to reduce greenhouse gas emissions. Much of this was drawn from the Dutch experience of attempting to set a cohesive, long-term agenda, reducing uncertainty for industry and creating the opportunity for both environmental improvement and developing new environmental industries and technologies (Wallace, 1995). A national initiative was established in the US through the Global Tomorrow Coalition's Sustainable America programme – a programme to create and document sustainable development strategies and encourage the incorporation of sustainability principles into decision making at all levels.

It has been argued that the impact of the PCSD and its initiatives was not as great as might have been expected, largely due to its perceived partisanship at a time of Republican ascendancy in Senate and the House of Representatives. The various initiatives in the USA have therefore gone ahead without any strong centralising focus on sustainable development, even though their aims are fully compatible with it (Bristow et al., 1997). At the local scale, the Metropolitan and Rural Strategies task force created four working groups on Rural Strategies, Innovative Economic Strategies for Sustainable Communities, the Multi-jurisdictional and Regional Collaboration working group and a group on Evaluating Progress.

The Rural Strategies working group addressed three questions. How can rural communities move from narrowly focused economic development to sustainable development? How can the relationship between urban and rural communities be leveraged to sustain rural communities? How best can the needs of rural communities be fulfilled? The Innovative Economic Strategies for Sustainable Communities working group identified how strategies that focus on people, places and markets can help build sustainable communities in metropolitan and rural areas. The Multi-jurisdictional and Regional Collaboration working group argued that a combination of community, government and market actors could form a powerful consensus, building a culture of inclusive engagement, to build sustainable communities. The working group recommended policies that provide incentives and dismantle disincentives and barriers to achieving multi-jurisdictional and regional co-operation on sustainable development in metropolitan and rural communities. Finally, the Evaluating Progress working group developed a national and community level framework of indicators of sustainable community development. At the national level, it was proposed that this framework could track national progress toward achieving the vision of sustainable communities established in Sustainable America. At the community level, this framework could recommend practicable performance-based measures that communities could use in setting goals and tracking and improving progress. However, the work of the PCSD came to an end with the election of George W. Bush and the appointment of an administration antipathetic to environmental issues. Given the new Bush administration's refusal to countenance signing the Kyoto agreement on climate change emissions it is unlikely that any support will be forthcoming at a national level for LA21 or other sustainability initiatives.

Progress on sustainable development over the next few years may thus depend on developments at local and state levels. At the state level, states have some autonomy on environmental issues, depending in part on their willingness and capacity to develop and implement federal legislation (Wallace, 1995). While some have been active in developing environmental policies, others have been less proactive and left local action to the EPA. In the former case, California, with its particular air quality problems in Southern California, has been especially active in relation to legislation on air pollution. Complaints from industry that this legislation was difficult to negotiate led in the early 1990s to attempts to radically overhaul Californian environmental legislation, in part to create jobs in the production of new environmental technologies to compensate for those high technology jobs being lost in defence industries. Activities in other states have included a conference organised by the state of Kentucky on state strategies for sustainable development, while Oregon's 'Benchmarks' programme has set goals and performance measures for sustainability. In Minnesota, the 'Milestones' programme defined 20 goals and 79 indicators to measure progress and the state's Sustainable Development Initiative has created sectoral development plans, while several other states have programmes on energy efficiency and renewable energy use (Corson, 1994). In addition, there are initiatives at the urban and local scale in the US on sustainability issues, some of which are outlined in Chapter 5.

In total though, despite some positive moves during the two Clinton administrations, both Democratic and Republican parties hold similar views on free trade and globalisation of the kind which place the environment a long way down the policy agenda. The US is unlikely to continue with even the limited national measures outlined above under the administration of George W. Bush and there remains a strong core of opinion within the US which is both anti-transnational and anti-environmental and which sees *any* global agreements as a threat to US sovereignty and the 'American way of life' (see Luke, 2000, for details). Much may therefore depend upon developments at the state and local level to implement measures designed to move towards sustainable development. Certainly, as with other nation-states, federal government has pushed environmental responsibilities down to subnational scales, albeit (as is also the case elsewhere) without providing the necessary resources (Whitfield and Hart, 2000).

Japan

Japan's history of environmental legislation is closely linked with its economic development. Rapid economic growth in the 1950s and 1960s led to both an increase in industrial pollution and a public outcry to curb resultant industrial excesses. Well-known cases, such as the Minamata mercury disaster, cadmium poisoning in Toyama prefecture and air pollution in Yokkaichi, led to litigation against companies and government legislation to control excessive air and water pollution. Between 1960 and 1992 Japanese industry invested US$75 billion in pollution control equipment and US$25 billion on energy efficiency measures (James, 1993). Such high levels of investment resulted in substantial reductions in industrial pollution and the erroneous view that Japan had conquered its air pollution problem (Nishimura, 1989). Moreover, it led to Japanese industry establishing a leading position in the manufacture and export of end-of-pipe environmental technologies such as flue gas desulphurisation and selective catalytic reduction for NO_x control (Longhurst et al., 1993).

The major local government bodies in Japan, the prefectures and metropolitan authorities, have considerable autonomy and have played a major role in balancing the needs of industry with local environmental concerns (Wallace, 1995). As at national level, the impact has been to encourage both technological progress and environmental improvement. Indeed, in many cases local legislative action has been ahead of national policy. Nationally, the introduction of a new Basic Environment Law in 1993 placed sustainable development as the basis of environmental policy. This had four main principles: recycling, coexistence, participation and international co-operation (Utsunomiya and Hase, 2000). As part of the Law, a Basic Environment Plan is required which obliges local authorities in areas with serious pollution problems to draw up an environmental pollution control programme (Wallace, 1995). Under this new structure, local authorities retain existing powers to set local environmental standards, but are limited in their ability to go beyond national standards on issues such as CO_2 emissions and other issues of global and national importance. The main impact of the Basic Plan has been to: encourage a shift towards voluntary measures by industry; develop a co-operative approach between industry and the broader community; encourage a move towards a long-term, comprehensive approach to environmental protection; and assist in the integration of economic and environmental policies. In June 1997 the Environmental Impact Assessment (EIA) Law was passed which requires all government projects and other designated projects (e.g. electric utilities, railway construction, forest roads) to produce an environmental impact assessment. Following the Earth Summit, the Japanese government began to create a National Agenda 21 for Japan and developed a government action plan involving the expenditure of 2.2 per cent of GDP each year in purchasing and contracts. This programme involves the government purchasing 'environmentally-sound products' to influence business and consumers and thus have a demonstration effect. Despite these national policies, development projects with adverse environmental impacts have gone ahead and pollution remains problematic – Japan is probably the most heavily dioxin-polluted country in the world (Utsunomiya and Hase, 2000).

Local authorities in Japan have taken much of the responsibility for environmental policy and have been instrumental in influencing the national policy context. Indeed, the 1993 Basic Environment Law requested local government to pursue environmental policy, mainly through planning and management of the environment. Subsequent decentralisation laws have allocated increased responsibility to the local level. In some cases, local government acted independently of national government prior to the 1993 Basic Law to produce their own 'environmental basic ordinances', effectively framework laws to cover all environmental policy. Following implementation of the Law, most prefectural governments and cities developed their own ordinances, even though this was not a statutory requirement. These are regarded as action plans, similar to Local Agenda 21 requirements (Utsunomiya and Hase, 2000).

At the national level, much of the development of policy in the late 1980s and early 1990s was a response by the ruling Liberal Democratic Party, and particularly former Prime Minister Takeshita, to criticisms that Japan was not playing an international role commensurate with its economic status (Falk, 1992; Coughlan, 1993). The environment was thus on the government's agenda due to top-down pressures and partly because it was a relatively non-controversial area for involvement, both internationally and domestically.[5] Japan emerged at this time as a major financier of overseas environmental programmes and an active player in international efforts to address global climate change problems. These activities were heavily influenced by the concerns and interests of the Japanese Ministry of International Trade and Industry (MITI) which saw commercial advantages to be gained (Schreurs, 1997). The result of government deliberations has been collaboration between industry and government to set stringent standards for emissions and to devise new technology to meet them (Gibbs and Longhurst, 1995). A number of joint government–private sector research programmes are intended to develop clean technologies to help achieve sustainable development. In 1990, more than 30 companies joined with MITI to provide US$76.9 million to set up the Research Institute for the Earth (RITE) for research into technologies to remove or recycle carbon dioxide and a number of other long term projects were launched (Box 3.5).

> *Box 3.5* Japanese policy initiatives on the environment
>
> - Launched a New Sunshine Programme, aimed at developing better anti-pollution technologies. This had a planned budget of US$14.9 billion over a 27-year period and intended to reduce Japan's energy usage by a third by 2030.
> - Launched a one hundred-year plan for the promotion of new environmental technologies and alternative energy sources, entitled New Earth 21.
> - Initiated a Green Aid Plan to assist developing nations in adopting environmentally friendly technologies. This programme, Technology Renaissance for the Environment and Energy (TREE) was designed to promote Japanese clean technologies in Pacific Asian countries.
> - Organised numerous regional and global conferences for the environment.
>
> *Source:* Adapted from *Far Eastern Economic Review* (1991); de Rosario (1992)

As Schreurs (1997) points out, this stance was in marked contrast to the late 1970s and early 1980s in Japan where, despite domestic successes in energy efficiency, recycling and air pollution control, there was only minimal attention paid to the environmental impact of Japan's economic development successes, a fact she ascribes to the lack of any substantial indigenous non-governmental organisation (NGO) community.

Despite the seeming innovativeness of such policies, the impact of stronger environmental policies within Japan has been limited. Proposals from Japan's Environment Agency for a tax to limit carbon dioxide emissions and to make environmental impact assessments (EIAs) obligatory for companies were watered down after protest from MITI and the Ministry of Construction, reflecting the weak position of the Environment Agency at that time *vis-à-vis* the more powerful economic ministries (Barret and Therival, 1991). The Environment Agency itself was almost abolished as a separate institution in the late 1990s, though after protest it was given ministerial status (Utsunomiya and Hase, 2000). Moreover, environmental concern does not always extend beyond Japan's own borders – this is evident in Japan's reluctance to control the environmentally destructive activities of Japanese companies overseas. For example, the 1997 EIA Law does not cover overseas development projects funded by the Japanese government (Utsunomiya and Hase, 2000). That

Japanese corporate overseas investment can be extremely environmentally destructive is well documented – for example, Mitsubishi Kasei, the chemicals group, was forced to close a plant in Malaysia after allegations of radioactive waste pollution (Terazono, 1994). Indeed, in recent years Japan has backed away from making specific environmental commitments, particularly for the reduction of carbon emissions. A severe economic downturn, coupled with the fragility of the banking sector and low consumer expenditure, has also combined to reduce national level interest in environmental policies.

Australia

Following the Brundtland Report (World Commission on Environment and Development, 1987), debate on sustainability in Australia revolved around the notion of 'ecologically sustainable development' (ESD). This represented an attempt by national government to regain control of the environmental agenda following a series of well-publicised clashes over forests, mining and wilderness (Christoff and Low, 2000). In 1989, the then Prime Minister, Bob Hawke, launched a Statement on the Environment entitled *Our Country, Our Future* and a national strategy for ESD was published in 1992 (Hawke, 1989; Commonwealth of Australia, 1992). ESD is defined as 'using, conserving and enhancing the community's resources so that ecological processes, on which life depends, are maintained and the total quality of life, now and in the future, can be increased' (quoted in Whittaker, 1997: 326). ESD was specifically seen as a means of emphasising the complementarity between economic growth and environmental protection, reflecting an approach similar to that of ecological modernisation (Bührs and Aplin, 1999). The commonwealth government produced an Ecological Sustainable Development Strategy for Australia and some state governments[6] developed their own plans. For example, ESD formed a key part of South Australia's Strategic Plan and the New South Wales government issued guidelines for ESD to all local councils. ESD and sustainable development featured prominently in Tasmania's Resource Management Planning System, Queensland's Environmental Protection Act (1994) and initially in Victoria's Sustainable Development Program for 1995–2000 (Whittaker, 1997). Despite this there is evidence that ESD is largely confined to environmental departments in some states. Thus in Queensland, 'while the conservation-oriented arms of the state government have embraced ESD concepts, there is little evidence that other departments, particularly economic or development departments, have incorporated ESD principles' (Hunt, 1992: 37). In Victoria a radically neo-liberal government subsequently instigated a 'degreening' of the state (Christoff, 1998). Indeed, from the mid-1990s the ESD strategy has been given a low priority by both national and state governments reflecting the changing economic priorities coincident upon the replacement of Bob Hawke by Paul Keating as Prime Minister and the election of the Howard coalition government in 1996 (Christoff and Low, 2000).

While the federal and state governments have adopted ESD, there is little reference to the local scale in its implementation. In early 1992 Australian governments at all scales (commonwealth, state and territory and local) signed an Inter-governmental Agreement on the Environment (IGAE) which placed sustainable development at the heart of policy and aimed at integrating economic and environmental considerations into decision making. This was also intended to solve some of the problems within Australian government relating to the division of responsibilities for environmental policy at different spatial and administrative scales. In Australia, local government largely depends on state governments for their legislative powers, as the commonwealth government has no direct powers over this level of government. Strategic planning and development schemes, which increasingly address local ecological issues, can be overridden by the states (Crowley, 1998). Typically it is at national and state levels that policy directives are set, although implementation is largely a local concern (Fowke and Prasad, 1996). At the local scale it has been argued that the development of LA21, and environmental policy more generally, in Australia has been constrained by the fragmented structure and decision-making processes of Australian government. As Wright comments:

> Australian local government is currently unable to achieve the goal of sustainable development. Local authorities are constrained from integrating social, economic and environmental policy at the local level due to the problems of inter-governmental structures and their own internal structures and decision-making processes.
> (Wright, 1995: 55)

Christoff and Low (2000: 253) argue that the IGAE merely served to reassert the powers of the states over national environmental policy and that in consequence the development of national policies on issues such as waste, water and air have been 'painfully slow', as have the development of local strategies.

Local government in Australia is therefore poorly resourced to deal with environmental concerns and fairly marginal to the process of Australian government (Crowley, 1998). This division of environmental powers between different levels of government has left local government with a narrow agenda including land-use planning, public nuisance, health and building controls. The ratification of the Rio Declaration and commitment to implement Agenda 21 led to proposals aimed at addressing these shortcomings in order to allow a more integrated strategic approach to issues of environmental management. Wright (1995) argues that this will necessitate local government adopting the approach of Integrated Local Area Planning (ILAP) constituted by: community consultation; developing information bases; preparing local strategic plans; and developing indicators to monitor progress. The Australian Local Government Association has also proposed that local government should play a key role in protecting the environment, integrating resource conservation and seeking balanced and sustainable development (Low Choy, 1992).

To date, however, progress in Australian local government appears to be slow. Research by Fowke and Prasad (1996) in New South Wales discovered that sustainability had a low priority in most respondent local governments and that few were preparing LA21 plans. A survey by Environs Australia in the late 1990s showed that only 33 out of Australia's 770 councils were submitting their policy processes to Agenda 21 criteria (cited in Christoff and Low, 2000). Even where these are in place, few encompass social or economic aims (Whittaker, 1997). Moreover, despite increasing environmental responsibilities, there has not been a commensurate increase in local government powers and funding. The main barriers to better environmental management in Australian local government, other than funding, are outlined in Box 3.6.

Despite these difficulties there have been a number of environmental programmes which, although these are national policies, have been developed with intergovernmental co-operation on a regional basis in the fields of land care, water management, regional development, biodiversity and coast care (Crowley, 1998). However, the most successful of these – Landcare[7] – has largely bypassed local government structures. The commonwealth government did introduce a number of measures to promote sustainable development involving local government including the National

---

Box 3.6 Barriers to better environmental management in Australian local government

- Poorly developed policy frameworks;
- Fragmented, out-of-date and ineffective regulatory systems;
- Inappropriate and ineffective intergovernmental relations;
- Lack of resources and skills at local government level;
- Local government income depends on a market-based property development system;
- Multiplicity of state and local approval systems with contradictory requirements.

Source: Brown *et al.* (1992)

Sustainability Program and training guides, but more recently national policy has shifted away from the ideals of ecologically sustainable development promoted in the early 1990s (Whittaker, 1997). While this began with the Keating government in late 1991, it gathered pace after the election of the conservative Howard coalition government in 1996. At the Kyoto summit on climate change in 1997 Australia opposed uniform global reduction targets in greenhouse gas emissions and eventually obtained agreement for an 8 per cent *increase* in emissions (Christoff and Low, 2000). In part this stance was due to changed national government, but also reflects Australia's economic dependence on primary commodity exports, such as coal and other minerals and the emphasis on rural areas as efficient sites of production in competitive world markets (Martin and Halpin, 1999). Whether ESD will have any substantive legacy remains to be seen, though Bührs and Aplin (1999: 325) argue that it may have led to 'tangible results via hundreds of more specific policies and programmes introduced in its wake at federal, state and local levels'. Given the nature of Australia's federal system of government, much more may rely upon state level activity than upon national policy developments.

The United Kingdom

In 1990 the UK government produced its first white paper on the environment entitled *This Common Inheritance* (HM Government, 1990), which was subsequently updated by annual reports. Boehmer-Christiansen and Murphy (1997) note two strengths of the document. First, that it was produced at all, given the then Conservative government's commitment to market forces and second, that it did represent a (small) step towards greening government policy and integrating environmental concerns into a wider range of policy areas. Despite this, the white paper was widely acknowledged to have major shortcomings, largely because it proposed very few new activities or measures for legislation, proposed no new public expenditure and lacked any quantifiable target setting. Following on from the white paper, the Conservative government produced a sustainable development plan, published as *Sustainable Development: The UK's Strategy* in 1994 (HM Government, 1994). This proposed some minor institutional changes, such as establishing a UK Round Table on Sustainable Development, but predominantly called on individuals, the voluntary sector and local government to 'deliver' sustainable development. In a similar fashion to the white paper, the Strategy was also criticised on the grounds that it lacked any target setting or objectives (Voisey and O'Riordan, 1997).

In the UK, a major stumbling block at the local level throughout the 1980s and 1990s was the limited engagement by central government with environmental policy measures, despite the rhetoric of the 1990 White paper *This Common Inheritance* and the subsequent UK strategy for sustainable development (HM Government, 1990, 1994). While central government envisaged local actors, and particularly local authorities, taking a key role in sustainable development policies, this remained largely in the realm of encouragement rather than providing the power and resources to implement policy. A report for the European Commission concluded that all of the most radical environmental activity undertaken by UK local authorities was non-statutory and, in contrast to the situation in some other member states, undertaken without financial support from central government (Expert Group on the Urban Environment, 1996b). Indeed, the increasing amount of environmental legislation from the European Union, and its supposed centrality for all areas of EU policy, offered local administrations the opportunity to either bypass or put pressure on, national government (S. Ward, 1996).

The election of a Labour government in 1997 led to the UK sustainable development strategy being revisited in the form of a consultation paper on the 1994 strategy itself and a renewed call for local actors to develop partnership approaches to Local Agenda 21 (Department of the Environment, Transport and the Regions (DETR), 1997b, 1998). The new strategy was less a strategy *per se* and more of a framework – as was its predecessor. Indicators were introduced, but again no targets were set other than aiming to move 'in the right direction'. Some local authorities had already seized on Local Agenda 21 as a means of legitimising renewed planning activity and environmental action, particularly as a reaction to the centralising tendencies of previous Conservative governments. However, much remained to be done and the Labour government wanted all authorities to adopt LA21 plans by the year 2000. Overall, though, despite initial enthusiasm for a 'greener' approach to national economic development, some of the key provisions, such as the development of an integrated transport policy, were soon shelved, either because of real or perceived external opposition or due to opposition from the Treasury. Proposals to shift taxation towards 'environmental bads', such as polluting vehicles or carbon usage, have either been minimal or shelved for further consideration. Other changes post-1997 have come about with the Labour government's devolution programme. There has been legislation to devolve government to Scotland, Wales and Greater London. The establishment of Regional Development Agencies (RDAs) in eight of the nine English regions from 1 April 1999, when they absorbed the existing bodies of the Rural Development Commission and English Partnerships, was the start of a key change in English regional governance. These developments have also been linked with the devolution of environmental responsibilities down to the subnational level. The Welsh Assembly has a statutory duty to promote sustainable development, while the English RDAs also have a responsibility for sustainability within their areas (Gibbs, 1998). Each of the English regions has had to prepare a regional economic strategy (with variable levels of commitment to an environmental agenda) and regional sustainable development frameworks. To complement these developments, planning legislation has altered to have a stronger regional focus. Regional planning bodies produce regional planning guidance, which must include a sustainability appraisal. Other national policy initiatives involve the creation of a Sustainable Development Commission to monitor progress on sustainable development (replacing the former Panel and Round Table on Sustainable Development) and developing official sustainable development indicators.

Local government, particularly UK local government, played a key role at the 1992 UN Earth Summit and have been in the forefront of developing Local Agenda 21 actions plans (O'Riordan and Voisey, 1997). However, despite the lead that the UK has shown, a third of local authorities did not have an LA21 strategy in place in 1997 (DETR, 1997b). The emphasis on the local delivery of policy in the UK can be seen in the central role ascribed to local authorities by both previous Conservative and the current Labour government. A recent call by central government to local authorities to renew their interest in LA21 justifies this by stressing (DETR, 1997b):

- The range of functions undertaken by local authorities, such as economic development, regeneration, housing and transport;
- The strategic position of local authorities in relation to their role as a catalyst in forming partnerships and acting as co-ordinators;
- The role of local authorities as major employers and consumers in their own right;
- The democratic accountability of local authorities and the legitimation this confers.

Box 3.7 shows the six main components of the LA21 process, as recommended by the UK government.

With regard to economic development strategies, local authorities are encouraged to develop policies that aim to provide 'fulfilling occupations' for local people, meet local needs and minimise environmental consumption. 'This may entail less emphasis on inward investment, more attention to nurturing local businesses, and treating Local Exchange Trading Schemes and community enterprises as core rather than fringe activities' (DETR, 1997b: 8). As Chapter 2 indicated, the extent to which this is happening has so far been limited with many more local areas being concerned with developing entrepreneurial strategies to align the local economy with national and international economic trends. In the absence of a supportive national framework it is difficult to see how local economies can make a decisive shift in this direction (Voisey and O'Riordan, 1997). Some indication that such a shift *may* be starting to occur is

> **Box 3.7** Components of a Local Agenda 21 process
>
> - Managing and improving the local authority's sustainability performance;
> - Integrating sustainability issues into the local authority's policies and activities;
> - Raising awareness and education;
> - Consulting and involving the wider community and general public;
> - Working in partnership with others – central government agencies, business, community groups and the general public;
> - Measuring, monitoring and reporting.
>
> *Source:* DETR (1997b)

provided within some of the regional economic strategies produced by the English RDAs where, for example, developing the environmental technology sector is a key theme, a theme that is encouraged by Department of Trade and Industry aims to create 'regional sustainable technology strategies'. The Department of Trade and Industry, long indifferent to the environmental agenda, developed its own (albeit tentative) sustainable development strategy in October 2000, with a particular focus on the need to encourage eco-efficiency within industry. Here the focus was effectively upon an ecological modernisation agenda, with an emphasis upon the role of innovation, business opportunities from resource efficiency and 'decoupling economic growth from unsustainable impacts on the environment and people' (Department of Trade and Industry, 2000: 6). A similar business-based agenda was proposed by the Prime Minister, Tony Blair, in a speech to a Confederation of British Industry/Green Alliance conference in October 2000, in which he emphasised the need for a more productive use of environmental resources and the opportunities offered by technological advance. However: 'there remains a palpable sense that New Labour is not really at ease with the environment . . . the inescapable feeling is that the environment does not really figure in the New Labour "project"' (Jacobs, 1999: 4).

Sweden

Sweden is usually seen as an environmentally progressive state and played a particularly active role in the Rio conference. Sweden was one of the first countries to implement some of the Agenda 21 recommendations from the

Earth Summit and recent shifts in government policy seem to indicate a desire to maintain global leadership in this area. In January 1997, the government created the Delegation for Ecologically Sustainable Development (DESD) within the Cabinet, consisting of the Ministers of Environment, Agriculture, Taxation, Basic Education and the Junior Minister of Labour (Lundqvist, 1997). The immediate remit of this group was to produce an integrative basis for government policy to create an 'ecological society', which took the form of a report – 'A Sustainable Sweden'. The measures proposed were:

- An annual sustainable development report should be produced using indicators to assess the environmental impact of all government spending;
- Developing an improved and revised set of environmental objectives for all sectors of society;
- A 'Greening of National Agencies' programme, utilising environmental criteria for public purchasing and auditing, as well as producing 'ecological sustainability assessments' of all major agency actions;
- A sustainability investment programme to run from 1998 to 2004, including 1 billion SKr for the ecological transformation of the built environment and infrastructure, 9 billion SKr to transform the Swedish energy system and 6 billion SKr for sustainability investments by municipal governments.

In the Cabinet's spring 1997 Economic Bill the DESD's proposed spending plans were adopted and around 12.6 billion SKr allocated to local investment programmes, infrastructure projects and for energy conversion schemes. A particular feature of these developments is a strong emphasis upon developing eco-efficiency through the use and development of new technologies (Lundqvist, 2000). A similar sign of increased government commitment is evident in the changed form of the annual Cabinet report on the State of the Swedish Environment. While this had previously focused on traditional environmental policy, from 1997 this changed radically to attempt the integration of environmental issues with resource management, economic and social concerns through action programmes. Finally, the Swedish government proposed a new Environmental Code to bring together existing environmentally relevant laws into one piece of legislation in order to make it 'more lucid, efficient and applicable' (Lidskog and Elander, 2000: 198). Increasingly, ecological issues in Sweden have been framed in the context of ecological modernisation ideals, where environmental sustainability and economic development reinforce one another (Lundqvist, 2000).

Part of the rationale for this shift in government policy was the economic and currency crisis faced by Sweden in the early 1990s with concomitant reductions in welfare benefits and social programmes and (by Swedish standards) high levels of unemployment. Lundqvist (1997) argues that the Social Democratic government promoted the development of an 'ecologically sustainable society' to promote new demands for new technologies and thus create a whole new ecological job market. Paradoxically though, the budget crisis also led to public spending cuts and cutbacks in the national state's own environmental institutions (Eckerberg and Forsberg, 1998). From 1997, the aim has been to combine environmental goals with labour market policy. In that year, an investment programme was launched to stimulate recycling within the building sector of 1 billion SKr, amounting to 30 per cent of total costs (Eckerberg and Forsberg, 1998).

Sweden has emphasised the importance of local level action from the outset of Agenda 21, with a guide for localities produced by the Ministry of Environment in 1994 and a National Committee on Agenda 21 established in 1995. Indeed, there has been an increasing transfer of environmental responsibility by national government to the subnational level in recent years. For example, in the 1993/94 Government Bill on the Rio Declaration, 'the municipality is acknowledged as having a key role for the *national* strategy geared at promoting sustainable development' (Lidskog and Elander, 2000: 206, emphasis in original). While much of the responsibility for action has been left to the local and regional scales, national government has acted as a source of advice and information about environmental goals and strategies, as well as allocating funding to local government and business for LA21 projects and ecological development. At the local scale, the municipal level of government is very strong in Sweden – all 288 Swedish municipalities had begun their own Local Agenda 21 by 1997 (Eckerberg and Forsberg, 1998). Local autonomy and tax-raising powers contribute to the development of local level initiatives, added to which the Swedish municipalities are required by law to incorporate environmental concerns into all their activities. Some of these are prescriptive, but in other areas municipalities are free to develop their own interpretations of national policies (Eckerberg, 1995). The Swedish Eco-Municipality Network, predominantly representing smaller and more rural municipalities, has assisted the spread of ideas about local sustainability strategies. Despite the seemingly supportive national framework in Sweden, Eckerberg and Forsberg (1998) point to survey work indicating that the municipalities consider national government has inhibited the development of strategies for sustainability. For example, tensions exist between municipal policies to implement sustainability and contrary tendencies in national policy, such as in energy, traffic and infrastructure (Lidskog and Elander, 2000). Despite leading the way in introducing carbon dioxide taxes in the 1990s, Swedish governments have subsequently backed off from imposing higher taxes on industry on the grounds of the impact on national economic competitiveness. Such tensions between economy and environment are also found at the local scale. However, Eckerberg and Forsberg contend that:

> most Swedish municipalities view LA21 not only as a strategy for sustainability, but also as a strategy for local economic development. This is particularly articulated in those municipalities with considerable structural problems, such as high unemployment rates. The support from local business is essential in two respects: both to extend the LA21 processes to include the supply side, and to legitimise those strategies that might otherwise be opposed by industry.
>
> (Eckerberg and Forsberg, 1998: 345)

Despite this Eckerberg and Forsberg conclude that locally, environmental policy is still largely subordinate to economic and employment policies. While it is possible to find good examples of local strategies based on sustainable development (some of which are outlined in Chapter 5), it is equally possible to find municipalities focusing on entrepreneurial strategies, or indeed, both at the same time. At the national level, it has been argued that though: 'environmental concern is more and more regarded as a vehicle for economic growth . . . in cases where the two goals seem to be in conflict with one another economic growth is still the heavyweight' (Lidskog and Elander, 2000: 213). Indeed, support from local business in LA21 is not always reflected at the national level where the Federation of Swedish Industry (Industiföbundet) has argued that national environmental legislation merely creates an uneven playing field. Instead, industry groups have argued that the Swedish government should focus its efforts upon influencing international legislation providing for equal conditions for industry at a global scale (Lidskog and Elander, 2000).

Conclusions

This chapter has examined the development of environmental policy during the late 1980s and 1990s. The evidence indicates that the Rio Earth Summit in 1992 provided a new catalyst for policy formation and implementation, even if in some cases this is in the realm of rhetoric rather than substantive action. There is a clear indication, however, that environmental policies have shifted away from a concern with issues of pollution and remedial action at the end-point of economic activities towards a greater concern with preventive policy and the need to integrate economic, social and economic objectives. The Brundtland Report and the Earth Summit in particular popularised the concept of sustainable development and encouraged its adoption as the basis of national government policies. At one level, the concept of sustainability has certainly entered into national and local policies and with it a greater interest in integrating economic development and the environment. This is not to be over-optimistic about the potential for change, nor to neglect the key role of concrete agreements to reduce greenhouse gas emissions which have so far eluded the international policy-making community and which received a major setback in 2001 with the refusal of the Bush administration in the USA to implement the Kyoto climate change agreement.

The shift to greater integrative activity is seen particularly within European Union policy where the intention (if indeed, not the reality to date) of environmental policy has been to encourage a shift towards sustainable development as the basis of all EU policy. Both the international scale and the supranational level of the EU still reflect the difficulties in making environmental policy central to intergovernmental agreements, let alone placing sustainable development at the centre of policy. In both cases, economic and trade initiatives (as outlined in Chapter 2) are prioritised as being of greater importance. Similarly, the sections on national policy developments indicate that within individual countries the shift towards integrating economic and environmental policies is proceeding at varying speeds and with varying degrees of importance. A common theme appears to be a tentative shift towards more collaborative and participatory policy making and a view that policy integration can have economic benefits in the form of technological innovation and the resultant creation of new economic sectors. Within Europe, in particular, integrative policies are also increasingly seen as having a key role in combating Europe's high unemployment levels. However, much depends on whether these policy initiatives can become mainstreamed as opposed to remaining tangential to what is seen as the 'real business' of economic development. As the evidence from Japan, Australia and US makes clear, the power of the environmental lobby *vis-à-vis* economic development interests in government and industry remains weak. In addition, as Chapter 2 made clear, at both international and national levels the influence of neo-liberal, free market economic policies remains strong and acts to counter much of the influence of policies oriented towards sustainable development. At times of economic crisis it is notable that national governments in the USA, Japan and Australia have backed off from implementing environmental policies on the basis that they will harm national economic competitiveness. This forms an interesting contrast with Sweden where, despite the usual debates and discourses over environment and economy, national government promoted a strong environmental agenda as a way *out* of economic problems through a process of ecological modernisation.

Notes

1 LIFE is the Financial Instrument for the Environment of the EU, specifically aimed at funding pilot environmental measures and projects. The Structural Funds are geared towards regional economic and social development activities and provide funding for projects and training activities. Most of the latter do not have a specific environmental focus, although there has been an increasing move towards incorporating environmental objectives into the output measures in order to assess environmental impacts.
2 These involve: cross-border co-operation, European energy networks and transnational development (INTERREG), adjustment to industrial change (RECHAR, RESIDER, RETEX, KONVER, the SMEs Initiative and ADAPT), local rural development (LEADER), the most remote regions (REGIS), urban policy (PESCA) and labour market integration (EMPLOYMENT).
3 These comprised programmes for development of lagging regions (Objective 1), conversion of old industrial regions (Objective 2) and development of rural areas (Objective 5b).
4 The European Commission, which is effectively the civil service of the EU, is divided into a number of Directorates-General. Each is headed by a Director-General and has responsibility for a specific area of EU policy.
5 As opposed to involvement in international military or peace-keeping operations, for example.
6 Australia has six states (New South Wales, South Australia, Western Australia, Victoria, Queensland and Tasmania) and two territorial governments (Northern Territory and Australian Capital Territory). In the text these are all referred to as state governments.
7 Originally entitled the National Land Management Program initiative, Landcare sought to bring together the various stakeholders in Australian land management to counter a long history of land and water degradation and to promote a more sustainable use of land, particularly agricultural land. It has been acclaimed as a success, both nationally and internationally, although doubts have been voiced about its longer-term viability and efficacy (see Ewing, 1996, for example).

4

INTEGRATING ECONOMIC DEVELOPMENT AND ENVIRONMENTAL STRATEGIES AT THE LOCAL LEVEL

Introduction

The theme of this book so far has been that while integrating economic development and environmental strategies is generally perceived as a desirable outcome by policy makers, there are a number of obstacles to actually implementing such strategies. In Chapter 2 the problematic nature of current trends in economic development for environmental policies was outlined. Local pro-growth strategies and the impact of globalisation, free markets and free trade have made it difficult for local strategies to encompass environmental aims. In Chapter 3 the emphasis turned to examining environmental and sustainability initiatives at the international, supranational and national levels and investigating the current context for local policy developments. While there have been important shifts at all these levels towards encouraging sustainable development and ecological modernisation, these are often fragile initiatives, liable to be abandoned in the face of short-term economic difficulties and seen as peripheral to the 'real' economy. Both Chapters 2 and 3 suggest that these changing policy frameworks make the development of strategies for sustainable development – encompassing environmental, social and economic aims – more problematic. However, while it is important not to downplay the sheer scale of change needed to effectively integrate these two seemingly disparate areas, if change *is* to occur then we need guidelines as to how this shift might start to come about. In this chapter I outline those guidelines that might help further develop such integrative policies and strategies. The chapter has a particular focus upon identifying these guidelines in terms of broad principles and themes, rather than outlining detailed initiatives and schemes – these form the subject of Chapter 5. In doing so, I am fully aware of the difficulties posed by developing such guidelines and subsequently actually implementing policies based upon them. Arriving at a clearer idea of what *needs* to be done does not, of course, guarantee that it *will* be done. However, while the task of implementation may be substantial and, as pointed out in Chapter 1, needs to be developed within a supportive international and national framework, I would argue that there *is* merit in developing general principles for sustainable development at the local scale, both for their value in assisting policy makers at the local level and as exemplars for policy at wider spatial scales. To paraphrase remarks attributed to Antonio Gramsci, we should possess 'optimism of the will' as well as 'pessimism of the intellect'.

Indeed, Ekins and Newby (1998) argue that such a focus on local economies has become *more* important in the face of globalisation, free markets and problematic national sustainability strategies. Their rationale for this is that:

- Local economic development provides opportunities for those excluded from participation in the global market. This may involve providing pathways to re-enter the global market or be productive outside it;
- Local economic development can maintain local economic networks and social coherence, both of which are increasingly lost in a globalised economy;
- Local economic development promotes local distinctiveness in the face of the cultural homogenisation of globalisation;
- A strong local economic base provides both stability in periods of restructuring and a base for new forms of comparative advantage to develop as opposed to the greater economic vulnerability created by globalisation.

They go on to argue that:

> [Local economies] ... often call on a commitment to locality which can be a strong motivating force in many people. They are informed by detailed local knowledge – of people, places and economic opportunities. And, perhaps most importantly, they are rich in human relationships, between producers and consumers as well as between producers and between consumers, which can inspire trust and commitment, effectively lowering transaction costs and facilitating the processes of economic interaction.
> (Ekins and Newby, 1998: 867–8)

While there is some truth in these arguments, I have already been critical of what I would regard as an over-simplified and over-romantic view of the benefits of more localised economies. As White and Whitney contend:

> there is a big contradiction here between two visions of the future – one based on international trade and the theory of comparative

advantage (whereby each country is supposed to produce those goods which it can competitively sell on the open world market), the other based on autarky (local self-sufficiency) which might be supported on environmental grounds as well as protectionist (employment protection) grounds.

(White and Whitney, 1992: 39–40)

A further criticism of these arguments might be that they appear to see local economic development as some form of parallel economy existing alongside the globalised economy. Thus Ekins and Newby (1998) see locally focused economic activity as drawing upon local assets and reserves, both human and environmental. The implication is that developing a parallel local economy based around local relationships and place-based businesses can provide employment and job satisfaction for those currently excluded from mainstream markets. While there is some justification for this as part of *any* strategy (in that jobs are provided and there may be an important associated intermediate labour market function), in order to make a decisive shift towards sustainability, change needs to be engendered in the current mainstream of economic activities. I would argue that while locally directed developments may be necessary, to see sustainable development and local economies in these terms is to continue the marginalisation of sustainability. Rather, sustainable development practices and principles should become the mainstream, albeit radically restructured, economy. This means engaging with existing economic activities within an area and, in particular, exploring the links between business and the environment. For example, Roberts (1995) addresses the interface between business and the environment to come up with a number of key themes that could form the basis for policy development. Much of this reflects the kind of political programme envisaged by the proponents of ecological modernisation outlined in Chapter 1.

The concern here, then, is to try and avoid both the over-optimistic scenarios of those taking an environmental perspective, who often make light of some of the difficulties involved (see Hines, 2000, for example), and the more pessimistic scenarios of those taking a predominantly economic perspective (see Korten, 1995, for example) to make some pragmatic suggestions as to how local and regional economies can be restructured. In broad terms, such an approach entails:

Working with the grain of the market where possible, but altering it through direct intervention where necessary. This side-steps the polar extremes of calls to overthrow the existing global order and the market-led economy on the one hand, or to rely almost solely on the market on the other hand.

(Haughton and Hunter, 1994: 202)

Thus, as a report for the European Commission commented:

> In the short term, much can be achieved through practical incremental steps in the right direction, seeking to reduce unsustainability as much as to achieve sustainability, by realigning existing policies and mechanisms and identifying a strong set of principles on which environmentally-sound action may be based.
> (European Commission, 1998b: 4)

In this chapter, following a brief reiteration of the fundamental principles involved in sustainable development, I then outline some general principles as a basis for implementing sustainability in local areas and regions. Having done this, the focus is then upon identifying those principles and themes that can specifically provide the basis for new forms of local and regional *economic* development. Finally, I outline how such principles and themes might come together in an overarching local or regional strategy, by reference to both 'deep green' perspectives, in the form of bioregional planning, and from the more technocratic, ecological modernisation-related perspective of regional environmental management systems.

Principles for sustainable development in local areas and regions

Haughton and Hunter (1994) provide a useful recap of the main foundation principles for sustainable development – inter-generational equity, social justice and transfrontier justice. However, the task is to try and find ways to operationalise sustainable development through moving from the broad principles of the concept to providing more detailed policy guidance of use for mechanisms of implementation (Bruff and Wood, 1995). Haughton and Hunter (1994) argue for a three-tier system of guidance, moving from these foundation principles, to guiding principles through to 'generally desirable' policy directions. This chapter focuses particularly upon the notion of guiding principles, while 'generally desirable' policy directions and specific initiatives form the basis of Chapter 5. Haughton and Hunter themselves have devised three sets of such guiding principles, based on ecological, socio-economic and management principles (see Table 4.1). Although developed for use in an urban context, these could equally be utilised for the development of any local area or region.

Other commentators have come up with similar lists of principles which they argue should form the basis of local and regional sustainability initiatives (see, for example, Artibise, 1995; Expert Group on the Urban Environment, 1996a; Mehra, 1997). Drawing upon these and especially upon the work of Haughton and Hunter outlined in Table 4.1, we might group these guiding principles into four main categories, though there is obviously overlap between them:

Table 4.1 Guiding principles for sustainable urban development

Ecological principles	Socio-economic principles	Management principles
• Prevention is better than cure • Nothing stands alone • Minimise waste • Maximise the use of renewable and recyclable materials • Maintain and enhance requisite variety • Identify and respect local, regional and global environmental tolerances	• Use of appropriate technology, materials and design • Create new indicators for economic and environmental wealth • Create new indicators for economic and environmental productivity • Establish acceptable minimum standards through regulatory control • Continue action to internalise environmental costs into the market • Ensure social acceptability of environmental policies • Widespread publication and participation	• Subsidiarity • Flexibility in devising and implementing environmental policy regimes • Long-term strategies are necessary for environmental management • Improved co-ordination across environment-related policies • Non-discrimination and equal right of hearing • Need for better availability and understanding of environmental information

Source: Adapted from Haughton and Hunter (1994).

- principles related to ecological factors
- principles related to resource efficiency factors
- principles related to governance issues
- principles related to the measurement of impacts.

Ecological principles

In relation to ecological principles, key ideas include the notion of limiting human impacts on the natural environment so as to remain within its carrying capacity. This is included, for example, in the British Columbia Round Table on the Environment and the Economy's principles for the foundation of sustainability (cited in Artibise, 1995). Associated with the idea of environmental limits, is that of applying the precautionary principle so as not to exceed the Earth's carrying capacity (Mehra, 1997). The European Commission (1998a) envisages a number of specific environmental measures where actions can be taken that contribute towards this including: improved air and water quality; protecting the built environment and green space; promoting resource efficient settlement patterns; reducing greenhouse gas emissions; and minimising environmental risks. Such measures may involve internalising environmental costs into the market with the intention of making users of environmental goods, who frequently currently receive these as 'free' goods, pay the full costs of their activities. This relates to concepts such as the 'polluter pays' and 'user pays' and has also been linked to attempts to shift the tax burden away from 'environmental goods', such as those associated with employment, to 'environmental bads', such as use of fossil fuels (Haughton and Hunter, 1994). Overall then there is a need to make local areas and regions more environmentally sustainable by methods which do not impose development costs on wider spatial areas (European Commission, 1998a).

Resource efficiency principles

Resource efficiency factors form another set of general principles to be considered. Here notions of demand management (managing demands rather than meeting demands) and environmental efficiency (reducing the use of natural resources and increasing durability) come into play (Mehra, 1997). Principles to be followed include: holding to a minimum the depletion of non-renewable resources and promoting long-term economic development that increases benefits from a given stock of resources without drawing on stocks of environmental assets. The European Commission argues that:

> Policies based on the principles of resource efficiency (optimising the use of material inputs and non-renewable natural resources per unit of output) and circularity (such as the recycling of materials, land and buildings) can lead to both reduced environmental impacts and costs savings, thus making economic as well as environmental sense.
>
> (European Commission, 1998a: 5)

Ideas around resource efficiency frequently invoke the utility of ecosystems management, where local areas are envisaged as complex systems characterised by continuous processes of change and development. Energy, natural resources and waste production are regarded as flows or chains. Maintaining, restoring, stimulating and closing the flows or chains can contribute to local and regional sustainable development. This is further developed through the idea of a dual network approach:

> The dual network approach is one example of an approach based on the principles of ecosystems thinking which provides a framework for urban development at regional or local level. This frame-

work consists of two networks: the hydrological network and the infrastructure network. The hydrological network defines ecological cohesion by managing water quantity and flows. The infrastructure network provides opportunities to minimise car mobility and to stimulate the use of public transport systems, walking and cycling.

(Expert Group on the Urban Environment, 1996a: 4)

In this sense then there is a need to learn from ecology in approaches to flow management, developing an integrated approach to closing the cycles of natural resources, energy and waste. The objectives here would be to minimise consumption of natural resources, especially non-renewable ones, minimise production of waste through reuse and recycling, minimise the pollution of air, soil and waters, and increase the proportion of natural areas and biodiversity in cities.

Governance principles

Principles related to governance issues stress the importance of having the appropriate structures in place to implement policy. The aim should be to provide a system of decision making and governance designed to address sustainability. The EU Expert Group on the Urban Environment (1996a) further develops these points in specific relation to urban areas. They propose that there are three overriding principles for sustainable urban management – co-operation and partnership, urban management and policy integration. These principles are very much about achieving a shift in governance structures and patterns, such that sustainability objectives are incorporated into existing local governance structures and there is an increase in democratic participation. Policies must be integrated both horizontally (that is across policy fields at local, national and international levels) and vertically (that is, between international, national and local policies). Co-operation and partnership approaches are intended to avoid individual organisations pursuing their own agendas, as well as being based on the argument that sustainable development can only be delivered through co-ordinated actions. Such partnership and co-operation may include consulting and involving the public, as well as creating partnerships (Local Government Management Board, 1994). Thus an important feature of this approach is the emphasis placed on involving communities in developing local action plans for sustainability (Jackson and Roberts, 1997). By this means it is intended that sustainability initiatives can contribute to both welfare efficiency (obtaining the greatest human benefit from each unit of economic activity) and equity (social solidarity and the equitable distribution of wealth). The principle is that of meeting basic needs and aiming for a fairer distribution of both the benefits and costs of resource use and environmental protection (Mehra, 1997). Finally, Haughton and Hunter (1994) argue that all these principles should be both socially acceptable and involve widespread public participation.

Some of the problems associated with following these governance principles can be illustrated from the example of the UK's 'Environment City' initiative. With reference to the partnership process involved in Leicester's Environment City, Newby and Bell (1996: 101) comment 'evidence suggests that while of great value in gaining initial participation, and in improving communications and co-ordination, partnership structures ... do not provide a magic recipe for cooking up a sustainable city'. They go on to argue that a successful set of first steps must be to:

- Establish a clear working definition of sustainability, together with a broad policy statement;
- Review the current state of the local environment and the local authority's impact upon it;
- Investigate the needs and priorities of local people;
- Develop detailed policies and strategies to achieve the overall aims, in line with expressed priorities;
- Establish a management system covering internal responsibilities, monitoring systems and future policy development to ensure that the programme remains on course.

Despite the best efforts of policy makers to proceed along such guidelines and to develop partnerships, Newby and Bell (1996) discovered that the involvement of private sector businesses was particularly low. Seminars, business networks and environmental reviews were all introduced and targeted at small and medium-sized firms (SMEs), but firms remained to be convinced of both the benefits of participation and their own environmental impacts. The way forward, they argue, is through: 'presenting local sustainability in clear and meaningful terms; forming partnerships; developing initiatives to maximise action within each sector; monitoring progress; and involving as many people as possible in planning and implementing change' (Newby and Bell, 1996: 102).

Carter and Darlow (1996) also examine the non-engagement of business in the context of the same 'Environment City' initiative. They argue that the Local Agenda 21 process is not seen as relevant by business compared to more immediate issues such as the state of the economy, that the longer term nature of LA21 goes against firms' short-term financial horizons and that there is a low level of awareness concerning initiatives such as LA21 and local environmental actions. To counter this they contend that private sector involvement can be increased by making use of umbrella organisations, establishing links with business at top executive levels and making the aims of environmental policy more directly relevant to business.

Measurement of impact principles

Finally, there is the principle that initiatives and policies must involve measuring, monitoring and reporting on progress towards sustainability (Local Government Management Board, 1994). This will mean developing appropriate sets of indicators. Here there is an increasing recognition that the measurement of well-being needs to move away from purely economic indicators such as GDP towards sets of indicators which recognise the interplay between the economic and the environmental, such as the Index of Sustainable Economic Welfare. At the local level, this may require an environmental audit in order to construct a baseline against which progress, as measured by the set of indicators, can be assessed. Related to this, Haughton and Hunter (1994) argue that we should introduce indicators of environmental productivity, such as water inputs per unit of output, to encourage the minimisation of resource inputs and waste outputs. These developments may be further encouraged by establishing minimum acceptable local standards, for example by establishing standards for water efficiency which encourage the development and adoption of new technologies.

Principles for sustainable local and regional economies

These general principles thus provide guidelines for the development of policies and initiatives intended to shift local areas and regions towards greater sustainability. However, they remain at a very general level of specification and, for the purposes of this book, do not provide policy makers with any clear suggestions as to how to integrate economic and environmental objectives. This section of the chapter therefore moves on to consider more specific principles for developing sustainable local and regional economies. At the most general level, there have been attempts which seek to outline broad principles for economic development. Box 4.1, for example, outlines those principles proposed by the UK Department of Environment, Transport and the Regions for a 'sustainable society'.

Even here, though, we have a set of very general principles where only the first and third points could be said to specifically relate to environmental sustainability. However, the case for local and regional strategies to integrate economic development and the environment has been made by several authors and in a number of policy documents. Indeed, Roberts (1995) makes a case for seeing the emergence of this theme as far back as the 1920s and the writings of Mackaye and Mumford, as expressed in their arguments for integrated planning and implementation of the economic development and environmental management of river basin regions. From their work, Roberts (1995) derives three fundamental tasks for a balance between economic development and the environment: the

> *Box 4.1* Characteristics of a 'sustainable society which promotes economic success'
>
> - A vibrant economy that gives access to satisfying and rewarding work without damaging the local, national or global environment;
> - Valuing unpaid work;
> - Providing necessary access to facilities, services, goods and other people in ways which make less use of the car and minimise impacts on the environment;
> - Making opportunities for culture, leisure and recreation available to all.
>
> *Source:* Department of Environment, Transport and the Regions (1997b)

conservation of natural resources; the control of commodity flows; and development of the environment.

Roberts goes on to argue that there are four broad principles which need to be addressed in attempting to redirect local areas and regions on to more environmentally sustainable paths:

- The application of concepts drawn from sustainable development which stress the environment, futurity, participation and equity. Thus considering the environment takes into account the full and true environmental costs of economic activities; while considering futurity maintains a minimum capital stock for future generations. Participation means that individuals share in decision making, while equity concerns provide for social justice at both intra- and inter-generational levels.
- A concern for the longer-term diversification and survival of a local or regional economy and developing the capacity to deal with future adversity. Policies should encourage diversification, full employment and a high level of local and regional control over capital and decision making.
- Increasing local self-sufficiency in order to minimise costly and wasteful resource inputs and transfers. Here there should be attempts to minimise the length and number of journeys and to plan the location of economic activity to integrate it with transport facilities. Energy efficiency should be maximised.
- Respecting territorial integration, both within and between localities and regions. This means localities and regions should develop a degree

of territorial integration as a unified space and ensuring that territorial integration rather than functional integration is regarded as the dominant force.

Here then we begin to move towards some more specific principles which can be used to develop policy and strategy, albeit still at a fairly general level of specification – for example, what exactly should territorial integration or local self-sufficiency mean in practice?

Indeed, Ekins and Newby (1998) address these concerns by arguing for a more holistic approach to what they term 'sustainable local economic development' (SLED) as encompassing both community development aspects and making existing business activity more environmentally sustainable. They argue:

> If any agenda for local economic development is to gain political support, it must be able to demonstrate not only a rich and positive impact on quality of life, but a significant impact in terms of job creation . . . a broad agenda for SLED must combine quality and quantity, and address both how to make existing economic activity more sustainable, and how to generate new enterprises and opportunities.
>
> (Ekins and Newby, 1998: 868)

Building upon this work and that of other authors including Roseland (1992), Local Government Management Board (1993), Roberts (1995), and the Expert Group on the Urban Environment (1996a) it is possible to outline the main principles and themes that should be considered when trying to develop sustainable local and regional economies (see also www.sustainability.org.uk). For the sake of clarity I divide these into three main groups: community and work-based themes, business themes and themes for local and regional government, reflecting the main groups of actors involved. As with the Forum for the Future's Local Economy programme (1998a) in the UK, the themes within this agenda are intended to form an interlocking set of initiatives, rather than being pursued in isolation.

Community and work-based themes

Capacity building and training: this would focus upon 'development from within', including community development work and training for self-confidence and motivation.

Community enterprise: involving greater access to capital by disadvantaged and community groups with a focus on those bypassed by the global economy. The aim would be to create benefits for the community as a whole rather than simply personal wealth. Community-oriented businesses can play an important role, both in terms of creating jobs directly and through community finance initiatives, such as credit unions and loan funds. Community economic development can be oriented towards controlling the local economy for narrow ends, such as increasing the capacity of a community to make money, or for broader purposes, such as increasing economic stability and control of resources, or to serve more fundamental goals of economic justice.

Access to sustainable and fulfilling work: this theme addresses issues surrounding the nature and distribution of work, with a strong emphasis upon work which makes a positive contribution to society and where individuals are assisted with improved access to work. Such 'sustainable employment' might include turning wastes into resources (e.g. recycling); improving efficiency with regard to energy and materials; converting to greater reliance on renewable energy sources; increasing community self-reliance (e.g. food and energy production); and sustainable management of natural resources (e.g. community forestry). Other initiatives could aim to increase employment through service sector activity or improved public transport.

Training and education: initiatives here can encompass providing training and education initiatives to meet company needs for a high quality workforce and help unemployed people find a job. Training initiatives can help to encourage employee awareness of environmental and sustainability objectives through environmental training.

Business themes

Responsible and responsive business development: this would encompass providing a supportive climate for new and existing businesses, improving environmental and social performance in existing businesses and developing new environmental industries. It would also include developing responsible business practices, including factors such as participatory decision-making processes and introducing environmental management systems.

Meeting local needs through local resources: here the emphasis would be upon linking local production with consumption (e.g. in sectors such as food, timber and energy), reducing transport use and pollution and increasing the 'local connectedness' of business, through for example encouraging local buyer-supplier chains.

Appropriate and clean technologies: a key issue is that of technological change, both in the sense of utilising more 'appropriate technology', but also in encouraging the use of technologies which work towards clean production, as envisaged in the ecological modernisation scenarios in Chapter 1. Technologies should first shift industry from materials-intensive, high throughput processes to processes which make more efficient use of raw materials and fuels, generate little or no waste, recycle residues and have inputs with low environmental costs. Second, 'technologies should aim to enable the use of nature's income, rather than consuming nature's capital' (Heaton *et al.*, 1991: 2). Technologies that increase energy efficiency within firms are also of key importance. Some authors have also argued for the 'decarbonization' of economies, i.e. a shift from carbon-based to hydrogen-based fuels (Hawken, 1987).

Ecosystems approaches to industry: this would involve thinking of firms within the local and regional economy as a potential linked system. An ecosystems approach to industry could be adopted by mapping resource flows at the local or regional scale, co-ordinating the development of industry sectors to maximise resource synergies, keeping the local economy as closed as possible, as well as providing sustainable infrastructure and encouraging business to locate near their workforces, customers and suppliers. Similar concepts have been developed in the context of demands for an ecosystems approach to urban development where cities are conceived of as complex systems characterised by flows as continuous processes of change and development. One approach to moving towards greater sustainability in economic development is that of industrial ecology. Industrial ecology essentially argues for a major restructuring of economic activities (especially in manufacturing industry). It has particular relevance for local and regional development, given that it envisages, amongst other things, the interchange of wastes and products amongst spatially adjacent firms. Industrial ecology approaches to sustainable local economic development rely upon utilising material and waste streams, so that manufacturing byproducts become the raw materials of subsequent processes leading to reduced or eliminated waste (Hawken, 1987).

Industrial ecology is based upon an analogy with natural ecological systems, the view being that firms can imitate ecological systems where organisms live and consume each others' waste. The focus is upon connecting different waste-producing processes, plants and industries in order to minimise the total amount of waste going to final disposal or lost in intermediate processes (Brand and de Bruijn, 1999). Effectively, the case is for 'dematerialization' of economies, using less in production but involving a greater amount of embodied knowledge in products – that is, a much greater input in terms of design, utility and durability.

Environmental auditing and assessment: using auditing and assessment to establish policies and management procedures.

Environment and accounting: adopting a long-term view which uses discount rates incorporating the real gains of enhanced environmental values.

Law and insurance: need to respect existing legal strategies and anticipate more rigorous legislation.

Packaging and product design: developing niches and sectoral specialisation. Consumer products would shift to becoming a service rather than a product with companies supplying goods that meet a need. Thus:

> Sustainability means that your service or product does not compete in the marketplace in terms of its superior image, power, speed, packaging, etc. Instead, your business must deliver clothing, objects, food or services to the customer in a way that reduces consumption, energy use, distribution costs, economic concentration, soil erosion, atmospheric pollution, and other forms of environmental damage.
>
> (Hawken, 1987: 139)

Marketing and public relations: using an enhanced environmental image for local and regional promotion.

Themes for local and regional government

Sustainable approaches to inward investment: this would involve a rethink about the aims and objectives involved in inward investment strategies, which are often a key feature of many local and regional economic strategies. Any inward investment should not take place at the expense of 'development from within' and should include an appraisal of the social and environmental impacts. Changing inward investment strategies may involve more selective approaches to investors, based upon their environmental performance and ability to link into broader strategic developments – for example, encouraging the environmental technology sector. Inward investment policies should be developed to selectively attract socially and environmentally responsible companies and those which strengthen local and regional 'business-to-business' trading relationships.

Design standards: this would encompass developing local design standards in areas such as energy efficiency, as well as in the durability and reparability of products.

Reuse developed and derelict land: already developed and derelict land should be reused. New industrial developments should involve landscaping and compensation where relevant.

Business support measures: this would include encouraging positive local supply side developments through providing advice, loans, infrastructure and support and demand side developments by informing consumers about, and encouraging, green consumerism. Such consumer preferences and local support should be built into policy. Local and regional governments can provide services for the support and development of SMEs, including support for new firms and community businesses. The aim should be to promote organic growth of the local economy, taking a proactive approach to business support and incorporating environmental good practice and social responsibility into this. Such measures would include encouraging responsible business practice, through the example of a local authority's own operations, selectively assisting firms which demonstrate this and encouraging firms to introduce positive approaches to working conditions. Government bodies should take advantage of new service and product opportunities to encourage new start-ups and focus inward investment.

Investment in the economy: this would involve encouraging public and quasi-public investment into sustainability or environmental funds. Where possible, these public and quasi-public bodies should be allowed to raise money in markets and invest in sustainability initiatives and enterprises. Such funds could also be utilised to attract and encourage environmentally aware business and environmental industries.

Co-operation and partnership: developing partnership arrangements with the local business community is a key feature of any strategy for local and regional sustainable economic development. There is a need to establish a culture of co-operation and partnership with the business community, considering business needs within strategy. Links between business and the research community should be encouraged to promote a flexible and creative economy. Such partnership working can help to create a local economic strategy based around sustainability and to argue for more environmentally sustainable policies at regional, national and supra-national levels. Developing peer group statements and local exemplars may also help to influence local business. Roberts (1994) argues that area-based partnerships should be developed to encourage both the 'greening of industry' and economic regeneration based on sustainable principles. In areas such as packaging, chemicals or the dissemination of environmental information, the concept of 'voluntary agreements' between the authorities and local industries has been particularly successful in Germany (where some 40 schemes exist), the Netherlands (more than 30) and in Belgium and Portugal (European Commission, 1995).

Transport policies: these should be integrated with inward investment strategies and with the needs of existing businesses and their workforces.

Integrating transport policies with inward investment strategies should ensure that businesses are able to meet transport needs with minimal environmental impacts.

Good practice: this would relate to local and regional authorities' own in-house practices and also to their own purchasing practices. Local purchasing policies should be developed which link firms to the authority's own purchasing strategy, as well as encouraging local 'business-to-business' trading and waste exchange networks, and encouraging local consumer purchasing.

Sector studies: environmental priorities and opportunities can be identified in local sectors to form the basis of transition strategies. Local self-reliance can be increased through a process of 'import-substitution' – though it can be argued that the simple substitution of products and services is merely the important catalyst to release local energy and inventiveness which in turn will encourage greater local self-reliance (Roseland, 1992).

Environmental quality: this would involve building local competitive advantage, as well as retaining existing economic activity by developing and promoting the environmental quality and quality of life in local authority areas. Addressing local pollution problems will also assist in improving the local environment and could provide the basis for future technology-based sector developments in some areas.

Implementing sustainable local and regional economic development: bioregionalism and regional environmental management systems

In this section of the chapter I consider how these processes of encouraging a shift to a more sustainable economy can be combined into an overarching strategy at the local or regional scale. One approach to this type of holistic planning is that of bioregionalism. Bioregionalism is a concept that is said to date back to the ideas of Lewis Mumford, Patrick Geddes and John Muir (McTaggart, 1993). 'In essence it involves the protection of the physical and ecological environments while promoting local cultural and economic stability' (Stevenson and Ball, 1998: 192). Definitions of a bioregion vary from simple watersheds to more complex definitions which include economic and social factors. Sale (1974) defines a bioregion as having boundaries which are determined by natural order, such as the attributes of flora, fauna, water, climate, soils and landforms, as opposed to human administrative and political dictates. While there is no agreed international definition, Miller (1996: 11) suggests that a bioregion comprises 'a geographic space that contains one whole or several nested ecosystems, characterised by its landforms, vegetative cover, human culture, and history, as identified by local communities, government agencies, and scientists'. Stevenson and Ball (1998) argue that for a bioregion to work successfully there needs to be a sound economic structure based on the area's resources. Bioregionalism would therefore place emphasis upon the need for each region to 'live within its natural limits to growth and that at the local level, social, economic and ecological factors are all synonymous' (Welford, 1995: 176). It is therefore potentially a force for radical structural change with a philosophy rooted in deep ecology and the aim of restoring a sense of place and community acting as a counter to the processes of globalisation (see, for example, Sale, 1985; Andruss et al., 1990; Snyder, 1990, for this viewpoint and Lewis, 1992, for a critique). From a geographical perspective, one critique of bioregionalism is that it appears to assume a causal link between space and society – remapping society into smaller, bioregional units will lead to a greater concern for the environment (Frenkel, 1994). The bioregional model emphasises local activity, local development and protecting the environment in a proactive manner:

> It stresses the importance of the local economy, of local employment and the development of local trading networks which are less reliant on traditional forms of mass transportation. The bioregional model also challenges the common acceptance of the need for large organisation, for large structures and for centralised government and institutions.
> (Welford, 1995: 183)

In practical terms, however, real life instances of bioregional policy have been less radical than all this suggests, although it can be argued that bioregionalism is simultaneously both a conservation and restoration strategy and a political movement for the devolution of power (McGinnis et al., 1999). Miller suggests a more pragmatic definition of bioregional planning as consisting of:

> An organisational process that enables people to work together, acquire information, think carefully about the potential and problems of their region, set goals and objectives, define activities, implement projects, take actions agreed upon by the community, evaluate progress and refine their approach.
> (Miller, 1996: 11)

In this sense then, bioregionalism becomes less a force for radical change and more of a methodological approach. Under this definition, there have been some attempts to utilise such bioregional planning concepts at the level of watersheds. For example, in Canada the Mackenzie Basin Impact Study has sought to investigate the implications of global climate change upon environmental, social and economic practices within the 1.8 million square kilometres of the Mackenzie river basin (Cohen, 1995). Indeed, such bioregional planning principles have been applied particularly to other rural and wilderness areas in order to conserve areas of biodiversity such as the Serengeti in Tanzania and Kenya and the Greater Yellowstone ecosystem in the USA, though this might be thought of as ecosystem management, rather than bioregionalism (Diffenderfer and Birch, 1997). Certainly there have been few, if any, attempts to apply the concept to urban areas or to seriously think through the implications for local economies, other than the type of optimistic post-Fordist scenarios criticised in Chapter 2. While the concept may therefore have potential application in the longer term, as a guide for the integration of economy and environment of the form envisaged here, it would appear to have limited applicability

More realistically, developing the themes and principles outlined in this chapter into an interlocking set of initiatives within a local area or region is initially likely to require a more technocratic and interventionist approach by governments at all levels. As Blowers points out in relation to planning:

> The essential characteristic . . . is that goals or outcomes are specified by the planning authority, be it local, state or international. In order for toe desired goal to be achieved, a planning process will need to identify targets, specify methods for achieving them and possess the capability for monitoring and evaluating outcomes and, where necessary, the sanctions necessary to secure compliance.
>
> (Blowers, 1994: 185)

One attempt to achieve this can be found within the EU's Fifth Environmental Action Programme, outlined in Chapter 3, which envisaged the development of 'regional conversion plans' within Europe involving the integration of economic development and the environment by means of a partnership approach between industry, the local community and local authorities. While the European Commission was rather vague about how this was to be achieved, some authors have proposed that such 'regional conversion plans' or 'regional environmental management systems' (REMS) could be implemented, drawing on environmental management systems approaches (Welford and Gouldson, 1993). It is argued by Welford and Gouldson that these REMS could be used as a marketing tool by a region or local area in order to project a good image and to attract in both environmentally-aware workers and companies. As Welford (1995: 175) comments: 'the underlying approach of the regional environmental management system is to develop a plan ... which will, over time, lead to the development of a comparative advantage based on integrated environmental management, at both company and regional level.'

In developing this plan, a code of conduct would be established through a partnership approach which would, amongst other things, devise emission targets, develop waste reduction strategies and encourage the adoption of environmental management systems in companies. This is essentially seen as a voluntary approach, based on co-operation and agreement between all partners in a local or regional economy. Obviously a key sector to involve is the private business community who need to be convinced of the benefits of such an approach for their own operations (Welford, 1995). As Haughton and Hunter argue:

> It may well be central to the success of urban environmental policies that representatives of major local private interests be included on any environment forum. This may in turn heighten the local business community's awareness of environmental issues, and improve their sense of ownership of the environmental problems towards which their activities contribute.
> (Haughton and Hunter, 1994: 229)

The idea here then is to apply the notion of private-public partnerships, which as Chapter 2 indicated have been a key feature of urban regeneration initiatives in many countries in recent years, to environmental problems. Welford envisages benefits for all from the introduction of a REMS, enhancing economic prospects by integrating economic, social and environmental factors in an integrated system. This then turns into a system where firms co-operate in dealing with pollution and other environmental problems, sharing and reducing costs. Certainly given the problematic involvement of the private sector in the Local Agenda 21 process referred to earlier in this chapter, the rewards to business will need to be tangible. The benefits to the local area or region may also accrue through 'branding' the locality as environmentally progressive with high threshold environmental criteria encouraging inward investment.

The development of a REMS would rely upon introducing some variant of an environmental management system which provides 'self-imposed, market-led' codes of practice to ensure quality and integrated environmental management. For firms, there would be a close link between internal environmental management systems (such as ISO 14000) and the REMS. This would also entail the adoption of a more holistic approach to environmental impact assessment, which could include the use of area-wide environmental audits, against which a range of policies could be assessed (Haughton and Hunter, 1994). Assessing regional progress would thus rely on establishing baseline data and undertaking regular monitoring to measure progress with the regional conversion plan. Much of the responsibility for the REMS would rest with some form of planning team with responsibility for a ten-point strategy (see Box 4.2). From this, a conversion plan would be developed for the region by means of a development committee comprised of local community and industry representatives to provide democratic legitimacy.

Obviously this would be a far from easy task and represents a major shift in thinking about local and regional economic development. Welford's

---

Box 4.2 Strategy to develop a Regional Environmental Management System

- Develop a regional environmental information system, including an environmental monitoring system;
- Conduct an extensive environmental and socio-economic study in the area, with emphasis on topics such as migration patterns, effects of infrastructural developments, employment, and the relative importance of the different economic sectors;
- Extend contacts and communication with local communities and industry;
- Develop internal strategies to promote information and education for all residents of the area regarding environmental protection and the benefits and implications of the REMS and to stimulate community involvement and environmental awareness;
- Design a template and model for processes and protocols within the area, resulting in improved communications and increased organisational control;
- Outline a regional environmental policy, with specific standards and targets;
- Conduct a detailed environmental quality survey to identify and prioritise areas for environmental rehabilitation and development;
- Stimulate and co-ordinate further development and implementation of company environmental management systems;
- Identify and prioritise areas for economic investment and growth, and identify new economic opportunities;
- Develop strategies to promote and market the area externally and to develop a scheme that gives recognition to 'green companies' within the region.

Source: Welford (1995)

own thoughts on how this would be achieved are somewhat optimistic. In tones reminiscent of Soviet planning, Welford envisages these REMS being implemented by:

> A REMS development team, with representatives from local communities and industries, and local government will set environmental targets and protocols at all levels. Targets will be continuously re-assessed and every person in the community will have their own environmental responsibility.
>
> (Welford, 1995: 179)

However, while this 'pure' form of a REMS may be improbable, others have argued that there is a real 'opportunity ... for the integration of company-based environmental management systems with the planning, development and management of local and regional areas' (Roberts, 1995: 214). In a very tentative fashion, this may be at an early stage in the operations of the UK's Regional Development Agencies, where regional economic strategies and regional sustainable development frameworks have been developed in tandem, and at a rather more advanced stage in the National Environmental Policy Plan of the Netherlands. Whether such REMS can be developed may depend upon the existence and development of what Marshall (1998) terms 'environmentally intelligent regional governance'. He argues that there are several factors that influence the extent to which this is present. First, there is a need for a strong regional jurisdiction to create effective sectoral policies and to ensure that these are integrated. In this sense then, Marshall's is arguing that modes of governance based upon networking and partnership are inadequate in themselves. Despite their increasing importance, they have limited negotiating power compared to strong institutional forms and 'institutional change can influence, over time, substantive outcomes' (Marshall, 1998: 439). Second, strong environmental sectoral planning assists with strong spatial planning. Third, the involvement of the economic arm of government is crucial in reforming strategies along ecological lines. Marshall points to the example of Lower Saxony in Germany where the Ministry of Economy, Technology and Transport has played a central role in transforming environmental sectors and established an Ecological Fund (DM 200 million between 1991 and 1994) to support the transformation of local industry. If such industrial or economic sectoral policy is weak, then Marshall argues that only dialogues and pilot initiatives can be introduced at the regional level, precluding the development of significant regional strategies. Fourth, the power balance between the private and public sectors is significant. Again, Marshall points to examples in Lower Saxony where private interests crowded out the public sector and argues that partnership always risks being too private sector-led. Fifth, reiterating the point made at the start of this chapter, the scope for local and regional action depends upon the framework established by central government. In particular the ability to regulate the private sector and the institutional and financial capacity to do so depends upon powers granted by the nation-state. Finally, there needs to be effective transmission from the regional to the subregional level.

One key area of concern with such developments is the need for policy integration across a number of areas, particularly for those local and regional government departments concerned with economic development, planning and the environment. REMS, or their local equivalent, would require maximum co-operation across departments, the integration of different departments' environmental policies and maximising the complementarity of environmental and other policies. From the arguments outlined in Chapter 2, it is apparent that this has become more problematic in recent years with the effective privatisation of many local government functions in countries such as the UK, as well as more general problems in co-ordinating across departments and functions. Such co-ordinating efforts are not impossible, but they will involve the integration of policies not just within local administrations, but also across a wide range of public, private and parastatal institutions.

Conclusions

This chapter has outlined the main principles and themes that need to form the basis of policy initiatives to develop sustainable local and regional economies. It is argued that these provide a starting-point from which to develop a set of interlocking initiatives that could substantially change the form and trajectory of economic development within a local area or region. However, such substantial change will not come about by a piecemeal approach. While it is possible to pick out some principles and themes from the lists given in this chapter and develop policy initiatives from these, at best such a strategy will only lead to minor changes. The point that is being made is that sustainability should be central to all aspects of economic development within a local area or region. Sustainability issues need to become a mainstream concern and not just confined to a number of isolated, if well-meaning, initiatives and projects. Of course, it is essential that we understand the development of such alternatives as representing a challenge to existing capitalist social relations and thus not just a simple matter of minor modification. To some degree their 'success' within this context may depend upon the extent of the challenge they offer. Some initiatives may offer only a minor challenge. For example as Eisenschitz and Gough (1993: 183) explain: 'community employment initiatives can pioneer new, risky but low capitalisation sectors which ultimately become part of the mainstream, as has been the case with health foods, green products and some ethnic minority ones.' On the other hand, more substantive challenges are unlikely to succeed. For example:

nation-states are constituted by their control over the money supply; this requires the national government to have exclusive control over bank credit, aggregate public spending and borrowing and, hence, interest-rate policy. No national government will tolerate local money, favoured by the Greens, on any significant scale.
(Eisenschitz and Gough, 1993: 183)

Again, then, we come back to the need for supportive national and international contexts for local level actions. Certainly what is not being proposed here is any notion of 'sustainable development in one locality or region'. While such actions may offer examples and useful experience, ultimately if environmental problems are to be addressed seriously then there needs to be radical change throughout national economies and in all areas. Otherwise there is potential for the displacement of environmental degradation elsewhere and the expansion of local areas' 'environmental footprints' (White and Whitney, 1992; Wackernagel and Rees, 1996). While therefore the argument in this book is for a local and regional approach to sustainable development, I would stress that this is ultimately envisaged as occurring within a broader spatial and administrative framework. As Peck and Tickell (1994b: 324) point out in a different context, there is a role for local solutions, but 'many of these will require regional, national and supra-national co-operation. Local strategies have a role, but this must be within a supportive and supra-national framework'.

This argument is neatly encapsulated by Jahn who states that:

> Activating the 'local' is now considered a *necessary*, though not *sufficient*, condition by environmental policy-makers. It is necessary because strategies to solve environmental problems, in order to be successful, need to be broken down to the experiential base of the local.... The local, however, is also not a *sufficient* condition for environmental policy-makers because environmental action can by no means be just local, it needs to be supplemented by regional initiatives (since urban fields span many jurisdictions and traditional distinctions such as the city and the countryside have become rather meaningless); it needs to be backed by those general and 'universalist' environmental policies not abandoned by the national 'Schumpeterian workfare state', and they need to be synchronised with global and transnational activities of environmental advocacy and policy-making.
> (Jahn, 1991: 54, emphasis in original)

What is being rejected in this book is a call for local autarky and any possibility of creating insulae in the world economy. While there is a sound case to be made for issues such as greater local self-sufficiency, greater local participation and linking local producers and consumers, the reality (at least on any time-scale realistically imaginable) is that local outcomes are increasingly influenced by distant events (Giddens, 1990). However, while not all the relevant activities can be undertaken at the local and regional level, it is far too easy to become overwhelmed by the sheer scale of environmental problems and overlook the potential for successive, and cumulative, small-scale actions which can help to resolve these problems. Such an incremental strategy may be able to achieve substantial results, albeit that it is not so dramatic an approach as calling for the introduction of bioregional and deep green solutions (Roberts, 1995).

Finally, it is essential that strategies are tailored to the particularities of local areas and regions. Returning to Haughton and Hunter's (1994) set of guiding principles for sustainable urban development outlined at the start of this chapter, there is a need for flexibility in devising and implementing environmental policy regimes at the local level. Thus:

> Complex problems demand complex solutions, especially given the variations in environmental tolerances and in government systems. Therefore rather than trying to utilise individual policy instruments derived from a particular philosophical approach or principle, it is more likely to be fruitful to consider a 'portfolio' of instruments (e.g. a combination of externality pricing, differential taxation, grants, subsidies and regulation) when responding to a set of interacting and changing environmental problems.
> (Haughton and Hunter, 1994: 227)

In Chapter 5 some of the initiatives that have been developed to try to make these small steps in particular areas and regions are outlined in some detail with a view to showing what is possible at these scales.

5

SUSTAINABILITY AND ECONOMIC REGENERATION
Making it happen on the ground

Introduction

Having outlined in Chapter 4 the general themes and principles that need to be considered when attempting to generate sustainable local and regional economies, this chapter turns to more specific examples of initiatives that have been developed in response to the sustainability agenda. Such lists of projects and policies are frequently used in the environmental literature as a means to celebrate and publicise individual initiatives, as well as indicating what can be achieved in often-difficult circumstances. There *is* room for the celebration of the possible – it can be argued that we need 'the self-conscious development of economic and ecological alternatives within this public sphere or "new commons" – alternatives such as green cities, pollution-free production, biologically diversified forms of silviculture and agriculture, and so on' (O'Connor, 1994: 172). However, while this chapter also provides details of particular projects and policies from a variety of locations, the intention here is to see these as potential elements of an overarching strategy for local and regional sustainable development. The argument is that such individual initiatives must be located within a broader framework of purposive policy action at the local and regional scales, as elements of a broader shift towards an economy based upon sustainable development. We have already explored the bases for such policy action in some detail within Chapter 4. In line with the basic tenets of sustainable development, it can be argued that such local and regional policy to integrate economic and environmental issues needs to have three main components (Roberts, 1995). These are:

- An economic component: the search for new forms of economic activity and types of business organisation that demonstrate a high degree of consideration for the environmental consequences of their operations and which attempt to minimise harmful impacts.

- A social component: the desirability of moving towards a pattern of spatial and social organisation that minimises the unnecessary or excessive use of resources, maximises the environmental benefits and which generates a higher overall standard of social welfare.
- An environmental component: the benefit of meshing together the business and societal elements with spatial concerns in order to provide for the environmentally responsible and socially balanced planning and development of localities and regions.

In parallel with the progression of work on principles and themes examined in Chapter 4, several initiatives which attempt to integrate these components have developed and there are some instances of innovative schemes and imaginative thinking. In the following sections of this chapter these are examined on a thematic basis. An attempt has been made to categorise these initiatives and projects on their major focus. However, there is inevitably an overlap between categories (for example, schemes can be intended to encourage the take-up of new technology, create jobs *and* be targeted at small firms) and it is arguable that some projects could be classified under different headings to those given. The chapter concludes with some thoughts on how policy makers can take these initiatives forward through a more strategic approach.

Land use and transport planning

This has a key role in organising the spatial relationships between economic development, infrastructure, population, markets and resources. Similarly transport policies and the development of related infrastructure play an important role in the relationships between places of residence and employment, as well as in the movement of goods to and from economic activities. Initiatives to do this can attempt to ensure that local transport strategies meet inward investors' transport needs, with a minimal impact on the environment. This could be through encouraging location near to good public transport services or locating close to rail or waterway transport in cases where companies have large freight transportation requirements. For example, the 'Business Circle' initiative in Vreten in Stockholm led to local companies influencing the public transport timetable in the area, as well as mapping out transport flows to and from business in the area, with the aim of encouraging joint, more efficient, journey planning (Forum for the Future, 1998b). In Trondheim, Norway the municipality introduced a fee for cars entering the city centre, with differentiated tolls by time of day and revenues used for transport investment. A similar scheme, but directed at air transport, has been introduced in Zürich with a landing fee surcharge imposed upon high emission aeroplanes (Erdmenger and Schreckenberger, 1998).

Advice and support for local business to improve environmental performance

While larger and multinational companies may have a key role in some localities, it has also been argued that 'it is unlikely that real progress can be made without the active participation of the majority of small and medium enterprises (SMEs) in partnerships for environmental stewardship' (Roberts, 1995: 243). However, many such companies lack the resources, expertise and confidence needed to develop an environmental improvement programme and may well lack any formal process of strategic planning. The aim here may be to encourage the 'organic growth' of the local economy, providing advice and training to local companies, targeting assistance towards new start-ups and the expansion of existing small companies and incorporating principles of environmental good practice and social responsibility into the criteria for such support programmes.

Various schemes to address the concerns of SMEs have been developed. For example, the Internal Environmental Care System in the Gelderland region of the Netherlands has established a formal environmental management system in small enterprises, including training for company managers, undertaking assessments of production processes and advising companies on environmental management procedures. In the London Borough of Sutton, a 'Business Eco Logic' project has provided consultancy on energy saving, waste management, resource use reduction and environmental management to local small and medium-sized firms. In Wiltshire, the Wiltshire Centre for Sustainable Development aims to provide education and training courses, as well as general advice and information and demonstrations of resource management and new environmental technologies. For business, the Centre's service will include business advice and support, a 'start up' facility for sustainable business activities and energy advice. In Bradford, the Bradford Business and Environment Support Team (BEST) was established by Bradford City Council in 1994 to provide SMEs with practical environmental support. It also ran the Bradford Business Environment Forum, which provided local businesses with information on environmental legislation and opportunities, encouraged environmental action and acted as a platform for disseminating good practice information. BEST was responsible for running the consultation process for Bradford's Business Local Agenda 21 and worked with a number of organisations to produce an environmental training package for use within the national network of Business Links, including Bradford City Challenge, RETEX (an EU programme aimed at regenerating textile producing areas), Bradford City Council, and the local Training and Enterprise Council.

In Lower Saxony the government set up an Ecology Fund of DM 280 million for the ecological restructuring of the economy. This has been especially designed for SMEs, although also supports communities and larger firms (Griefahn, 1994). The fund supports projects that:

- Apply and use new and renewable forms of energy and energy-saving ideas;
- Avoid, reduce and recycle waste and residual materials;
- Develop and test innovative and environmentally friendly products and processes;
- Provide specific environmental counselling and education;
- Stimulate investments in environmentally and socially compatible tourism.

Various local authorities have encouraged the development of 'green' business clubs or networks. For example, Calderdale and Kirklees Green Business Network in the UK aims to encourage good environmental management in local business and through this to improve competitiveness, efficiency and employment prospects. The network offers practical support and advice to local business, especially small and medium-sized firms, runs a waste exchange project and packaging waste compliance scheme, as well as encouraging interchange of experience through networking meetings (Forum for the Future, 1998b). Advice is provided in areas such as pollution, waste minimisation, energy and water efficiency and on environmental legislation and management systems. The exchange of information and experience on improving SME environmental performance and linking this to improved competitiveness has also been the subject of an EU collaborative project involving Bristol, Berlin, Copenhagen and Vienna (see Mills, 1997, for details).

In Sweden, the concept of 'Business Circles' has been developed whereby firms come together of their own accord to address common environmental problems. In the Vreten district of Stockholm in Sweden, a Business Circle has been established between thirty local businesses with the aim of firms working together to improve their environmental performance in conjunction with the Local Agenda 21 group and a waste management company. The various initiatives developed by the Vreten Business Circle include joint waste reduction measures, measures to increase energy and water savings, the environmental appraisal of buildings and research into commuter transport (Forum for the Future, 1998b). As with other schemes, it is estimated that participating firms have not only improved their environmental performance, but they have also improved their competitiveness and reduced costs (waste management costs at some firms have been reduced by as much as 60 per cent). The co-operative nature of the Business Circle reduces the cost to any individual firm of researching possible actions, has highlighted the potential for bulk purchase and shared deliveries and makes recycling more feasible on a common basis reducing costs and allowing transport co-ordination. Indeed, some of the actions and benefits in such schemes resemble the concepts involved in the development of eco-industrial parks outlined later in this chapter. As with many initiatives, it is said to be difficult to quantify the benefits in terms of profits or jobs. However, the fact that this initiative is (unusually) led by private sector firms and that they have remained participants would seem to indicate that there are positive benefits.

The extent of involvement by business and by property owners in this Swedish example stands in contrast to the situation in the UK where even in the high profile 'Environment City' of Leicester, there has been only limited interest and participation by the private sector as outlined in Chapter 4 (Newby and Bell, 1996). As Forum for the Future comment:

> Business motivation to take environmental action may be higher in Sweden than in the UK, perhaps reflecting greater consumer pressures (and hence promotional benefit) from taking action. However, in Sweden many companies have gained hard financial rewards that go beyond the more intangible benefits like image or acting responsibly.
>
> (Forum for the Future, 1998b: 24)

The main challenge in other countries may be to convince business of the benefits to be gained and find 'business champions' to take on this role and to work in partnership with other companies. There is still much work to be done within SMEs to promote the argument that sustainability will be a key component of business competitiveness in the future and that the so-called 'triple bottom line' of environment, economy and social issues is vital for future business operation. Finally, as a part of such initiatives there is also scope to encourage responsible business practice in firms, such as increasing local job accessibility and flexible working hours, whether through simple encouragement or, more practically, through linking assistance to improved practices.

Using technology for sustainability

The issues surrounding technological change and sustainability can be broadly divided into two main areas. First, there has been an argument that the trend towards greater use of information and communication technologies will have positive impacts upon the environment and contribute to sustainability. Second, there has been a focus on shifts in production technology away from 'end-of-pipe' solutions to pollution problems and towards the development of cleaner technologies that reduce emissions as an integral part of the production process.

The potential offered through greater use of information and communication technologies (ICTs) was examined briefly in Chapter 2. The growth of dispersed and decentralised industry enabled through greater ICT use, involving greater local control over the economy, is a theme that has been frequently developed in the environmental literature (see, for example, World Commission on Environment and Development, 1987). In the process, more self-contained local economies will supposedly develop whereby local control somehow results in a reduction in resource use and pollution (Dauncey, 1986). The proposed benefits also relate to changing work practices, in particular the growth of teleworking which, it is often proposed, can not only reduce pollution caused by commuting, but also help to bring productive work back into local economies (Robertson, 1986). Encouraging teleworking has been the subject of a number of pilot projects in some US states including California, Oregon and Washington (see Chapter 2 for details).

Similarly, teleconferencing, telebanking and teleservices have all been proposed as having potential benefit in reducing car-based travel (Friends of the Earth, 1997). Another theme is the notion that such developments allow the creation and/or retention of jobs and the provision of services in more remote areas, help to prevent out-migration, provide enhanced public sector services, such as health services, and encourage dispersed economic development. In Chapter 2 the example of Klamath Falls in the state of Oregon was outlined, which has involved a programme of distance learning, teleworking and telemedicine facilities in order to achieve such benefits. Similar initiatives on a much smaller scale are found in the UK. For example, in the London Borough of Sutton a telematics group has sought to develop computer-based technology and information networks for sustainability. This has included developing an on-line business database for networking and local purchasing, giving public access to environmental information and promoting home working. Many rural areas in the UK and the rest of Europe have experimented with 'telecottages' and 'wired communities' in attempts to combine job creation and retention, economic development and maintain social cohesion. However, as indicated in Chapter 2, the proposed beneficial environmental effects from greater ICT usage are not always forthcoming. Indeed, there may also be a number of negative environmental effects such as those associated with increased car-borne travel consequent upon the growth of out-of-town teleservice firms.

In relation to the development of clean technologies, there have also been a number of initiatives. One of the most ambitious has been the redevelopment of the Emscher 'corridor' in Nordrhein-Westfalen in Germany. This is an 80 km area suffering the environmental effects of one of the largest concentrations of traditional heavy industry in Europe. In the corridor, a partnership of institutions and economic operators with long-term environmental objectives have, with European Union funding, pursued a strategy of rehabilitating disused industrial sites. The Emscher Park site has many examples of the application of environmental technology and has acted as a demonstration project to disseminate experience gained on the environmental dimension of regional development (www.iclei.org/egpis/egpc-o39.htm).

Support for environmental technologies at the local scale has been a particular feature of developments in other parts of Europe. In Berlin, the Environmental Improvement Programme for small and medium-sized firms offers subsidies of up to 50 per cent of the cost of environmentally sound investments and technologies. This programme aims to reduce pollution, disseminate new technologies and innovative solutions and raise environmental awareness in SMEs (www.eaue.de/winuwd/50.htm). The programme is estimated to have achieved significant environmental improvements in waste, energy and emissions reductions while at the same time creating jobs. In Cork, Ireland, the Clean Technology Centre provides research, information and educational material on clean technologies to subscriber companies and carries out pollution abatement trials at company premises. In Lower Saxony, the government has piloted a scheme to allow the introduction of ecological control of all corporate activities in firms – a more sophisticated combination of an environmental audit and the introduction of an environmental management system (Griefahn, 1994). In Copenhagen, environmental technology and urban renewal have been combined through the development of an urban ecotechnology centre in the old covered market in the Den Brune Kødby district (European Commission, 1995).

Promoting green consumerism and purchasing policies

The objective of these initiatives is to enable both corporate and individual consumers to identify and to value products produced more sustainably so as to give them market advantage. Providing the relevant information to consumers can have an impact on behaviour. Examples include the provision of consumer advice by Lothian Regional Council in Scotland, in conjunction with Friends of the Earth Scotland. Local authority purchasing and tendering can also have an impact on corporate behaviour, given that such administrations are large consumers of goods and services. Local firms can be given the opportunity to bid for local authority contracts, as well as encouraging local 'business-to-business' trading, waste exchange networks and other initiatives to encourage the retention of local generated wealth within the locality and improve company resource use efficiency (www.sustainability.org).

Local authorities can also provide information about local producers to encourage the concept of local purchasing by both firms and individual consumers. In the UK, North East Lincolnshire Council has developed a 'Buy Local' project to encourage local small and medium-sized firms to expand their markets and put them in touch with larger local and national companies (Forum for the Future, 1998b). This has involved 'Meet the Buyer' evenings and developing a Buy Local Net Bulletin Board where companies can advertise their need for a good or service and local companies can respond as potential suppliers. More recently, the whole 'Buy Local' project is accessible through a computer network. Although not initially conceived of as having any specific environmental benefits and promoted on the basis of business efficiency, the scheme encourages the retention of wealth within the locality and has also made firms aware of the need to introduce environmental and management quality schemes if they wish to capture part of the larger firm market.

Similarly, in St Paul, Minnesota, the 'Home-Grown Economy' project experimented with a number of attempts to establish closed loop, self-sustaining networks. For example, rather than dispose of tyres to landfill, experiments were conducted with freezing them in liquid nitrogen, pulverising them and then using them as a filler for repairing potholes in roads (Roseland, 1992). In Germany, the government of Lower Saxony has developed an environmental directive to encourage the use of environmentally friendly products and promote 'ecologically advantageous proposals from those seeking building contracts and contracts for other supplies and services' (Griefahn, 1994: 425). Criteria include reducing water consumption, reducing emissions, avoiding waste and using recycled products, as well as criteria for new buildings related to issues such as heat insulation, use of passive solar energy and collection and use of rain water. In the London Borough of Sutton, the council has tried to ensure that all the companies it works with have an environmental policy and, ideally, a recognised Environmental Management System in place.

Targeted inward investment strategy

Some local areas have targeted inward investment on the basis of the potential contribution towards an improved environmental performance. This has implications not only for marketing, but also for strategic and physical planning and for the provision of sites and premises, infrastructure and financial support. Such strategies involve creating the conditions which will attract new companies to the area, such as a good quality environment, but in addition encouraging the relocation of companies which are socially and environmentally responsible and those which strengthen local and regional trading relationships. Examples include the proposed development of a 'sustainable business park' in Freiburg, Germany, intended for businesses devoted to having a positive environmental effect.

Such developments have also been proposed as a means of developing new industrial estates based on environmental principles. However,

Brand and de Bruijn (1999) argue that actually implementing such initiatives is problematic if the plan is to interlink companies through the provision of joint facilities and through the interchange of waste products and energy flows. They argue that public policy has to rely on the coincidental supply of investors and that few public sector agencies have the competence to design such systems of interchange. A more realistic option may be to encourage greater environmental awareness on an individual basis by companies. An example of this is the Dunkerque Industrial Environment Planning Scheme in Northern France, which aims to renew Dunkerque's image as a site for industry and to encourage new investment at the same time as maintaining and improving the local environment (www.sustainability.org). This aimed to create a consensus-derived planning process where new investors agree to environmental 'ground rules' for each sector (such as limits on air emissions, waste and energy conservation measures and recycling targets), the introduction of landscape management principles to protect the local environment, monitoring the impact of industrial development though environmental and health surveys and developing 'bio-indicators'. New companies sign up to a 'charter' committing them to the Scheme's requirements. Far from such (albeit voluntary) regulation constraining industry, it is argued that close involvement in drawing up and implementing such a scheme has increased industrial commitment to reducing emissions and resource use. The more problematic element comes with those firms already located in the area prior to the scheme and which remain a focus for local environmental concern.

Changing governance structures

Policies developed around attempting to introduce new governance structures can be seen, in part, as attempts to implement Welford and Gouldson's (1993) notion of Regional Environmental Management Systems (REMS) outlined in Chapter 4 – a locality or region-wide plan for environmental improvement, agreed between public and private agencies, which can provide a source of comparative advantage for both companies and the area. Again, an example is the development of the Emscher Park scheme in the state of Nordrhein-Westfalen in Germany to promote structural change in the Emscher region, an area of industrial dereliction in the Ruhr valley consisting of a corridor 80 km in length. The main objective of this scheme has been the ecological renewal of the area to create a new basis for economic development. The intention has been to develop business or science parks which specifically aim to encourage environmentally friendly businesses or firms specialising in environmental technologies, but which also make provision for other uses, such as housing and child care.

In the UK, the Kent Prospects Sustainable Business Partnership (SBP) aims to establish Kent as a centre of excellence for sustainable business development. Although led by Kent County Council, the scheme involves a wide-ranging partnership between business, local councils, public sector and voluntary bodies. The SBP developed out of a mix of existing local environmental initiatives and a more conventional economic development strategy developed for the county (www.sustainability.org). An outline action plan identified a number of priorities including: environmental management and promotion, optimising resources, optimising local production and purchasing, environmental innovation and sustainable development. When implemented, the plan involved a mix of existing projects, best practice demonstrations, promotions and research and development. The argument has been that by bringing projects together under the SBP heading, the action plan raises the profile of existing work, brings added value to it and links isolated projects together. Such schemes include an environmental awards scheme for business, a waste minimisation demonstration project and a tourism demonstration project to encourage resource efficiency and energy efficiency advice. Future plans include projects to encourage organic food production, supply chain initiatives and a centre of excellence for integrated resource management.

In Scotland, Jackson and Roberts (1997) chart the changes that have occurred in the Fife economy with a shift from a local economy based upon coal to military production and fabrication work for oil exploration and argue that these shifting patterns of dependence have had adverse impacts, notably persistently high levels of unemployment. In attempts to move away from an over-dependence on a limited number of sectors, the Fife Structure Plan produced by Fife Regional Council, is said to represent a proto-ecological modernisation phase of environmental management. This has involved three main elements: piloting an eco-management and audit scheme (EMAS) for the authority's own operations in order to focus on management processes dealing with the environment; introducing new management structures to allow greater scope for the implementation of LA21 and the EMAS scheme; and piloting the introduction of sustainability indicators as part of LA21. One intention has been to create a new path of development based, in part, around sustainability objectives.

Similar attempts to capitalise upon environmental advantage can be found in North America. Artibise (1995) outlines cross-national initiatives to deal with economic and environmental issues in the 'Cascadia' region of North America (covering the territory around Vancouver-Seattle-Portland) – these have effectively attempted to introduce new governance systems to deal with environmental concerns and represent a form of bioregional management for strategic planning. He argues that this is a search for new approaches to the management of the environment due to 'broad recognition . . . that the traditional structures of government are no longer effective in responding to the challenges of an increasingly interdependent world' (Artibise, 1995: 243).

In the UK the advent of Regional Development Agencies (RDAs) in the English regions from April 1999 has led to the emergence of sustainability issues in relation to regional development. Although sustainable development forms part of the RDAs' responsibilities, the level of interest in, and engagement with, sustainability has varied widely (Gibbs, 1998). Investigation of the RDAs' economic strategies in just three examples, revealed that engagement with environmental issues and sustainable development varies considerably. In Yorkshire and the Humber, there is thus much rhetoric in the regional strategy about sustainability – the argument is put forward that the RDA believes that the only way to achieve lasting growth is to create integrated, sustainable development – but the strategy is rather short on how this will be achieved or exactly what it means for the region (Yorkshire Forward, 1999). In the East of England Development Agency (EEDA) none of the actions planned for the first year of operation addressed any form of environmental or sustainability agenda. This lack of engagement with the sustainability agenda is perhaps surprising given that sustainable development was the subject of one of the EEDA's ten thematic working groups. This group's report argued that sustainability needs to underpin the regional economic development strategy and that this should occur through the integration of a set of sustainable development principles into the work of other working groups. However, it would appear that this has not fed through into subsequent strategy documents (East of England Development Agency, 1999a and b).

In North West England, the theme of the environment, particularly in relation to pollution and environmental liabilities, is one that runs throughout the strategy. 'Investing in the Environment' forms one of four themes within the regional economic strategy, which has two main objectives: restore and manage the region's highest quality environmental assets; and regenerate areas of dereliction and poor environment. These are developed through eight principles that run throughout the strategy – sustainable development being one of these. Sustainable development is said to be central to the strategy, even though the strategy states that it will involve hard choices where decisions will need to be taken balancing economic and employment objectives against environmental measures. Some innovative ideas are in the strategy – for example, much is made of the need for differentiated regional fiscal measures, although the likelihood of national government giving up fiscal control is remote, as indicated in Chapter 4. It is argued that the proceeds from such regional taxation transfers could help to fund programmes of environmental management, greening and land reclamation (North West Development Agency, 1999).

A broader attempt to shift the basis of development within a subregion onto a more sustainable basis can also be seen in attempts by Cumbria

County Council in the UK to conduct a sustainability appraisal of economic development within the county (Forum for the Future, 1998b). Some of the recommendations from the appraisal in 1998 were subsequently incorporated into the county's economy strategy. These included issues such as:

- Greater use of reclaimed materials and energy efficiency measures in the refurbishment of existing buildings in urban regeneration schemes;
- Incorporation of environmental awareness training into skills and learning programmes used by the local workforce;
- Integrating environmental business support with more conventional business support services;
- Minimising the impact of new premises on natural resources and the local and global environment;
- Integrating biodiversity aims into site reclamation schemes and encouraging business support for biodiversity action plans.

It is argued that these and other measures represent a considerable advance on previous economic strategies, which paid little attention to the environmental or social consequences of such development (Forum for the Future, 1998b). While there would appear to have been real attempts in a number of locations to adopt a more holistic approach to sustainable economic development and regeneration, there is little evidence provided of the relative importance attached to such initiatives compared to more conventional economic development measures, nor of how, if any, attempt was made to measure the impact of the sustainability appraisal.

Promoting the environmental business sector

This can encompass a wide range of initiatives, from low-skill, labour-intensive recycling schemes through to high technology-based clean technology development. Local economic development agencies and programmes can target support mechanisms, such as providing premises, infrastructure, grants, loans and advice. Examples include Lothian and Edinburgh Environmental Partnership (LEEP), founded by Edinburgh City Council to develop sustainability businesses in the fields of energy, transport and recycling. In Denmark, the 'Triangle Region' of Southern Denmark, encompassing eight municipalities, has a 'green city network' to foster commercial activities based on clean technologies and resource saving. Local authorities and the private sector have collaborated to pioneer approaches to waste treatment and combined heat and power, thus reducing energy consumption and emissions. Also in Denmark, 'Green City Denmark' is a partnership between four municipalities in central Jutland – Herning, Ikast, Videbaek and Silkeborg. In partnership with private companies, the aim is to provide a 'shop window' for Danish expertise in environmental technologies.

Other more substantive initiatives have seen the development of whole areas as centres for environmental industries and environmental technologies, for example, the 'Green Tech Centre' in Vesoul, Hautes Saône and the Porte des Alpes technology park near Lyon, both in France (www.france.net.au/official/cst/scient/fst27/fst27p4.html). These concentrate on encouraging inward investment from firms in sectors such as renewable energy, waste disposal, packaging engineering, noise reduction techniques and pollution monitoring. In both cases the development of such parks has relied heavily on basing the development around pre-existing companies within the local area and 're-branding' a combination of old and new developments around an environmental industries theme. A more recent development along these lines has occurred in Turin with the development of the European Environmental Science and Technology Park.

The Turin Environment Park claims to be the first European Science and Technology Park entirely dedicated to environmental technologies. It is located in the city of Turin in the so called 'Spina 3' area, the redevelopment of which is supported by Turin City Council. The park was conceived as a vehicle for the integration of environmental measures in production, consumer processes and re-imaging of the area, through the advancement of research and technology transfer to SMEs. Public and private research laboratories will be located in the Park. The Park offers services to SMEs both directly and by encouraging collaboration between enterprises and the academic world. The Environment Park is said to represent an ideal location to foster technological innovation and scientific research due to its vicinity to higher education institutes and advanced research centres. The initiative aims at encouraging the creation of innovative enterprises in the sector of environmental protection and sustainable development.

The Environment Park is designed to:

- Supply assistance and services to enterprises in environmental matters and to certify products and processes;
- Supply assistance to enterprises for eco-reconversion of production cycles;
- Develop environmental research on the basis of local expertise;
- Supply information and international contacts concerning technological evolution, environmental regulations and market dynamics in the EU;
- Organise scientific and technological training in specific sectors;
- Supply services to the public administration.

The Environment Park covers about 22,500 square metres where offices and laboratories are located in landscaped parkland, once a large industrial area. The buildings – laboratories, offices, service centres – are intended to set an example of environmentally compatible and sustainable architecture and energy conservation. The Environment Park project is also intended to contribute to improving the image of Turin and Piedmont in Europe and abroad, and constitutes the first major step in the forthcoming transformation of a vast unused industrial area in the north of Turin.

Companies from Europe and outside Europe, operating in advanced research and application of new environmental technologies are expected to locate in the Park. At present, the Environment Park hosts twenty-five enterprises. New landscaping will characterise the area. The inner core will match this 'green architecture' theme: all systems and plants in the Park building are designed to minimise the impact on the environment. Natural energy-saving systems, extensive use of renewable energy sources and eco-compatible, recyclable materials are the principles that have inspired the design project (Box 5.1). While the development of such a park may indicate that this is a viable proposition, the Environment Park has received substantial support, in financial and promotional terms, from the local city council, the regional council, the European Union and an extensive partnership of higher education and business interests.

Linking economic development, labour markets and social policy

Local labour market initiatives on recruitment and training can be linked to environmental goals. In many cases, such environment-related initiatives take on the role of intermediate labour market schemes, designed to move people into the formal labour market by a combination of training, personal development and work experience. Increasingly there has been a move away from a view that environmental measures cost jobs, towards the notion of 'win-win' situations, where employment can be created in conjunction with environmental benefits and economic development.

A number of schemes around the world are designed to take advantage of such 'double dividends'. In San Jose in the USA, a city plan was developed to create 170 jobs over ten years with an initial investment of US$645,000 which focused upon energy use and efficiency schemes and reducing energy use on government buildings and transportation (Roseland, 1992). In Amsterdam, energy teams offer advice and practical assistance to residents on energy-saving methods. In a four-year period, around 40 jobs were created, visiting 7,000 homes. Similar policies have created jobs in Glasgow through the Wise Group. In Berlin, the Senate launched the Berlin Ecological Renovation Programme for East Berlin in 1991 to meet both environmental and labour market objectives. This scheme co-funded employment and training initiatives in the environmental field and investments directed towards environmental protection and urban renewal. Also

> *Box 5.1* Activities in Turin Environment Park
>
> *Environmental certification*
> Establishing a committee to develop an observatory for the diffusion and the evolution of regulations concerning environmental certification, both for products (eco-label) and production sites and organisations (EMAS, ISO 14000).
>
> *Indoor and electromagnetic pollution*
> Creation, with the Polytechnic of Turin, of a Scientific Committee to develop an observatory on the initiatives of the sector as well as to support research and application works.
>
> *Renewable energy sources*
> The Environment Park facilities will encourage the application and development of technologies that use vegetal biomass to generate thermal energy.
>
> *Planning and management of water resources*
> The Environment Park has started to co-operate with various bodies working in the following fields: planning of water resources; management of water treatment plants; chemical analysis of water; optimisation of water resources used in production cycles.
>
> *Acoustic pollution*
> In collaboration with the Polytechnic of Turin and two companies located in the Park, a workgroup is being formed to study activities and initiatives in this field. A Promotional Committee bringing together different parties will be set up to carry out applied research.
>
> *Bioreclamation*
> A laboratory dedicated to bioreclamation and eco-toxicological tests to develop methods for environmental risk evaluation in relation to ground and surface waters and soil contamination.
>
> *Land reclamation*
> The Environment Park promotes a network of environmental companies based in several European countries working in this field through meetings and telematic links. This networking activity aims to support companies in sharing information, finding partnerships and funding opportunities.
>
> *Environmental software*
> As far as initiatives linked to computer science technologies are concerned, a workgroup is being formed to study the development of specialised software for the environmental sectors.
>
> *Telematics*
> The Environment Park participates in an experimental project called Torino-2000 for the use of high-speed fibre optic cables in co-operation with Telecom Italia and CSELT. Environmental tele-training through telematics for schools and experiments on teleworking is currently being studied.
>
> *Source:* Adapted from www.envipark.com

in Germany, the ZAUG GmbH (Giessen Centre for Employment and the Environment) was established to identify opportunities for employment creation in areas such as environmental improvement and protection and also to provide training for local unemployed people in these fields. ZAUG is a non-profit making vocational training company, directed primarily at young people, people on income support and the long-term unemployed. Funding for the company comes primarily through a local programme 'Work Instead of Income Support' administered through the local authority (www.sustainability.org.uk). The German benefits system encourages such action by local authorities, which have responsibility for the payment of unemployment benefit to the short-term unemployed within their own areas. ZAUG has two main areas of operation:

- Strategic services: including vocational advice and training for female returners to work, as well as apprenticeship training for young people.
- Business/enterprise units: including a mushroom farm, leech cultivation, an organic farm, household appliances and 'valuable waste' recycling and repair, regenerative technologies (solar energy, water reuse) and an environmental advice service.

On Merseyside in the UK the CREATE project (Community Recycling Enterprise and Training for Employment) is aimed at long-term unemployed people, providing them with training and skills, through salaried employment in the reuse and recycling of domestic appliances. It is argued that the project has had benefits in providing real work experience for the trainees with consequent improved success in gaining employment after the end of the training period, as well as environmental and social benefits in recycling domestic appliances and the provision or sale of these to needy households (Forum for the Future, 1998b).

In Sweden, the motives for developing LA21 initiatives in some of the 'leading' municipalities (Kugsör, Sala and Trollhättan) were to solve local problems with high unemployment and structural changes within industry. 'Local strategies for sustainable development have thus been a way to market the municipalities, that is, to develop a green image and thereby attract investors ... A related motive is the use of LA21 as a way to create new employment' (Eckerberg and Forsberg, 1998: 341). Cost-intensive environmental projects in these municipalities has been justified by reference to:

- Marketing potential: an example of this is the Sala Eco Centre, a centre for developing and selling know-how on sustainable development to industry and others. The municipality has invested SKr 3.8 million in the centre.
- Employment potential: several projects have developed on the basis of national government subsidies to combat unemployment or develop 'green workplaces'.

- Financial potential: where national funding availability creates the opportunity to develop projects at the local level.

Labour market policy can also be linked to the development of community enterprises to meet economic, social and environmental needs and/or establishing local exchange and trading schemes (LETS), which provide a means for people in a local area to exchange goods and services through the mechanism of a local currency. In the UK, the number of community initiatives like LETS and credit unions has recently grown substantially. At the end of 1992 there were 40 LETS in the UK, by 1997 there were over 400 schemes involving over 30,000 participants. The number of credit unions increased from 50 in 1986 to 520 in 1996 to over 600 by 2001 – which can also act as a means to link social and environmental policies (Forum for the Future, 1998b). Credit unions are financial co-operatives that offer low-cost financial services to their members and can be particularly beneficial to those on low incomes or who are excluded from mainstream financial institutions (Local Government Association, 1999). Credit unions not only encourage savings and offer loans to those who may not be able to provide security for a bank loan, they also prevent money leaking out of local communities and allow more local control of local money. They can thus play an important role in addressing community development, anti-poverty and sustainable development aims. Some credit unions, such as those organised through the Lothian and Edinburgh Environmental Partnership (LEEP) in Scotland combine small-scale lending with environmental objectives through the provision of loans for low-energy light bulbs and the purchase of season tickets for public transport. Other work by LEEP has focused on the potential energy savings to be made (and resultant lower fuel bills) from replacing old and inefficient domestic appliances in lower income households.

Compared to other countries, credit union development in Britain has been relatively weak. The reasons for this encompass restrictive national legislation, assumptions that credit unions should be small scale, voluntary and act as poor people's banks and the lack of one central credit union support structure in Britain (Local Government Association, 1999). This situation has begun to change slowly with the emergence of the Association of British Credit Unions Ltd as a major national body and greater recognition by local authorities of the wider role that credit unions can play.

Partnerships between local government and industry

One of the key relationships which needs to be addressed is that between local government and business within the local area. As Roberts comments:

to expect business to change its attitudes and operations in the absence of public policy inducement and regulation is naïve and unrealistic; whilst to expect governments to be able to implement change in the absence of support from business is to indulge in a degree of fantasy that has little foundation in terms of the realities of late twentieth-century economic control and management.
(Roberts, 1995: 243)

The need for partnership between business and local government is not just a reflection of the dominant discourse in regeneration strategy, as outlined in Chapter 2, but also reflects the inherent weakness of local authorities acting in isolation. While international and national factors largely determine local economic situations, working in partnership with other local and regional agencies may increase local effectiveness. A key factor here may be the need to ensure adequate business representation on partnership and other decision-making bodies and developing a successful dialogue with local business on sustainability issues. Effective partnership at the local scale can also assist in the development of local and regional lobbying for financial resources and/or changed fiscal regimes at broader spatial scales. This is particularly relevant for those regions within the European Union where EU regional funding has been of increasing importance. Again, several examples of partnership initiatives can be identified.

In Bilbao, action has been taken jointly by local government, other public agencies, universities and the private sector to revitalise the city. The local economy is based on sectors such as steel, metals, chemicals and energy which are characterised by high levels of air pollution which is exacerbated by the concentration of industry along a narrow river valley. The revitalisation process has been broader than a purely environmental focus, but it has also recognised that environmental degradation is a severe problem for the city and one which has a negative impact upon economic development. This is said to have:

> Engaged the active involvement and support of several sectors of local industry and helped to create markets for a fledgling environmental management industry in the region. More importantly, environmental issues are increasingly viewed not as a peripheral distraction, but as part of normal business management.
> (Expert Group on the Urban Environment, 1994: 122)

In Bologna, a Sustainable Production Programme has attempted to develop a partnership between local business and the local public sector bodies to encourage a restructuring of production activity by local firms, especially small and medium-sized firms (Bianconi *et al.*, 1998). A particular focus was upon encouraging moves towards cleaner production technologies by such firms. In Gothenburg, the municipal government established the Gothenburg Environment Project (GEP) to provide practical demonstrations of economy–environment integration, for example through developing environmentally friendly detergents to reduce pollution in Gothenburg harbour.

Provision of sustainability infrastructure

As the section on land use and transport planning earlier in this chapter indicated, providing sustainable transport infrastructure is one area where action can be taken at the local level, but other action can be taken on other forms of physical infrastructure. These could include the provision of heat distribution pipes, building office and business units to high standards of energy efficiency or providing facilities for environmentally efficient waste management, such as separation and sorting plants and incinerators with energy recovery. In Odense, Denmark, for example, industrial sites have been zoned so as to provide economic and environmental benefits, particularly through the use of waste heat from a combined heat and power station. Companies and residents thus gain inexpensive heating, while enterprises using surplus heat, such as market gardening, have also been attracted.

In Bergsjön, a district of Gothenburg in Sweden, a fundamental shift in the development of the area took place from 1993 onwards to convert the district into an 'ecological municipality' (Forum for the Future, 1998b). This has involved a number of large-scale projects and initiatives based around an ecological regeneration approach. This has included the refurbishment of housing in the district, including environmental measures such as energy efficient ventilation and solar energy, but also improvements to safety and the physical appearance of apartment buildings. Other schemes include the revitalisation of a drained marsh into a wetland area which has not only acted to clean surface water in the area, but has also provided jobs and training, the development of a low resource use community centre and related educational programmes, as well as the development of ecological enterprises and a recycling scheme. While the shift in regeneration policy towards the ecological municipality may have had environmental benefits, an assessment of the Bergsjön initiative concludes that it has had only a marginal impact on unemployment rates and other economic problems, despite improving the local quality of life.

In the UK one area of infrastructure provision has come through the construction of an increasing number of buildings designed to maximise energy efficiency. These include: John Menzies headquarters, Edinburgh; the Ionica office, Cambridge; Barclaycard's headquarters, Northampton; BRE's Office of the Future, Garston, Herts.; the Helicon Building, London; and Thames Tower, London. Another example is the Doxford business park on a greenfield site near Sunderland, which has as its flagship project the Solar Building designed by Akeler Developments. This building uses a combination of photovoltaic (PV) cells, building orientation, ground-coupling and air venting to control the temperature and to maximise the availability of daylight and fresh air. A south-facing atrium combines a façade of thin film PV glass (with movable solar blinds behind) with a stack effect to draw fresh air through to the offices behind the façade. Power is generated, but overheating – one of the biggest problems for office buildings – is also tackled. The building has attracted EU energy funds – reportedly of key importance given that electricity savings were not enough to offset costs. While this would seem to be an interesting model for future office development, it is worth noting that the Solar Building had not been let to tenants at the time of writing.

These private sector initiatives have parallels in construction initiatives with local authority involvement. Although never built, Hyndburn Council in Lancashire had well-advanced plans for a new 'zero-energy' office which combined high insulation and heavy building mass with an east–west orientation, a linear 'lightwell' for natural daylight and ventilation, and wind, solar and water power. It was also intended to connect up with the UK's national cycle network. It would have emitted no carbon dioxide, compared with annual emissions of 52 tonnes by a similar sized air-conditioned building. Similarly, the Eco Centre in Jarrow, opened in late 1996, was designed to be wholly autonomous in energy and services – it is currently estimated to be 90–95 per cent on target. It is powered by a wind turbine and solar panels, while sunlight, body heat and office equipment supply warmth. Much of the building is recycled – the roof is made from aluminium drink cans. Rainwater and 'greywater' from sinks and basins are also recycled and it is probably the first office block in the UK to use compost toilets. The building is naturally ventilated – a 'solar chimney' above the atrium creates a current of fresh air. Computer simulations have been used to orient it towards the sun, and green planting on exterior trelliswork or 'biotecture' allows shading in summer, but maximum natural light in winter when the leaves are bare. This building received national recognition from the Royal Institute of Chartered Surveyors, as the winner of its Efficient Building Award for 1997 and was built by Groundwork South Tyneside. The Eco Centre is said to be the first commercial building in Europe to take into account all aspects of the environment in its construction.

Industrial ecology and eco-industrial parks

A recent policy development has been an increased interest in eco-industrial estates or parks, based on ideas drawn from industrial ecology. The aim is to increase business success while reducing pollution and waste.

Proximity allows firms to exchange wastes to provide inputs to each others' processes and to make use of common environmental services. An eco-industrial park has been defined as:

> a community of manufacturing and service businesses seeking enhanced environmental and economic performance through collaboration in managing environmental and resource issues including energy, water, and materials. By working together, the community of businesses seeks a collective benefit that is greater than the sum of the individual benefits each company would realise if it optimised its individual performance only ... The goal of an EIP is to improve economic performance of the participating companies while minimising their environmental impact.
> (Lowe and Warren, 1996: 7.8)

The most often-cited example is that of Kalundborg in Denmark. A web of waste and energy exchanges occurs here between the city, a power plant, a refinery, a fish farm, a biotechnical plant and a wallboard manufacturer (Brand and de Bruijn, 1999). This operates as follows:

> The power company pipes residual steam to the refinery and, in exchange, receives refinery gas (which used to be flared as waste). The power plant burns the refinery gas to generate electricity and steam. It also sends excess steam to a fish farm, the city and a biotechnical plant that makes pharmaceuticals. Sludge from the fish farm and the pharmaceutical processes become fertilisers for nearby farms. Further, a cement company uses fly ash from the power plant, while gypsum produced by the power plant's desulphurisation process goes to a company that produces gypsum wallboard.
> (Brand and de Bruijn, 1999: 224)

The Kalundborg example has developed over a long time period and as a result of voluntary co-operation between the companies involved.

> The participants in the industrial ecology in Kalundborg believe that a co-operation like theirs can only develop if good relationships exist between the companies and the authorities. According to them 'creating industrial ecology is more a question of psychology than technology'. The possible partners need to be convinced of the economic advantages. The authorities can not regulate the process because often the options for exchanging waste in the region will not be visible to them. There is a role for them in stimulating initiatives by providing information; initiating a

discussion with industry to convince them of the economic advantages; identifying the persons who can be the leaders in the process; and to ensure that there are no laws that constrain the application of industrial ecology.

(Brand and de Bruijn, 1999: 225)

Attempts have been made to replicate the Kalundborg example elsewhere. For example, in the Rijnmond area, near Rotterdam, eighty companies set up an association with the goal of reducing waste and emissions from their industrial activities. There are different options for establishing industrial ecology-type experiments elsewhere, depending on whether policy is dealing with existing or new developments. Brand and de Bruijn (1999) outline a number of options for each case in the context of industrial estate development (Table 5.1). Similar 'industrial ecosystem' parks are being developed in Canada, Namibia, Fiji, the Netherlands and Indonesia. A particular interest has been in the USA, where the President's Council on Sustainable Development (PCSD) formed an active task force on eco-industrial parks as one element in building a sustainable economy. The US Environmental Protection Agency (EPA) and the Department of Energy have taken up these ideas to encourage the development of demonstration sites.

There is a broad spectrum of such eco-industrial parks (EIPs) in the US illustrating that there is more than one way to conceptualise and implement the concept. Current parks include:

- Brownsville Eco-Industrial Park, Brownsville, Texas
- Burnside Eco-Industrial Park, Nova Scotia (Canada)
- Civano Industrial Eco Park, Tucson, Arizona
- East Bay Eco-Industrial Park, San Francisco Bay, California
- Fairfield Ecological Industrial Park, Baltimore, Maryland

Table 5.1 Factors involved in developing an industrial ecology approach to industrial estates

Revitalising existing estates	*Developing new estates*
• Facilitating the management of waste streams	• Encouraging alternative energy sources
• Stimulating the building of energy and water cascades	• Stimulating the development of second water work facilities
• Efficient use of space	• Acquisition strategy
• Stimulating the creation of joint facilities	• Efficient use of space
• Mobility management	• Alternative transport systems

Source: Adapted from Brand and de Bruijn (1999).

- Franklin County Eco-Industrial Park, Youngsville, North Carolina
- Green Institute, Minneapolis, Minnesota
- Plattsburgh Eco-Industrial Park, New York
- Port of Cape Charles Sustainable Technologies Industrial Park, Eastville, Northampton County, Virginia
- Raymond Green Eco-Industrial Park, Raymond, Washington
- Riverside Eco-Park, Burlington, Vermont
- Skagitt County Environmental Industrial Park, Skagitt County, Washington
- Shady Side Eco-Business Park, Shady Side, Maryland
- Stonyfield Londonderry Eco-Industrial Park, Londonderry, New Hampshire
- Trenton Eco-Industrial Complex, Trenton, New Jersey
- Volunteer Site, Chattanooga, Tennessee.

Four of these were designated as demonstration sites by the PCSD in 1994 – at Baltimore, Maryland, Cape Charles, Virginia, Brownsville, Texas and Chattanooga, Tennessee – and these are perhaps the most developed. The stated goals for almost all of the parks include job creation, increasing the tax base, or other economic objectives. Four EIPs had education as one of their explicit goals. Some parks are visionary, with goals such as becoming the first multi-modal EIP with an ISO 14000 environmental management system in the US, becoming a zero-emissions or closed-loop manufacturing EIP, having all major tenants producing sustainable products with sustainable manufacturing practice, and becoming totally energy independent of fossil fuels or outside electricity.

The physical features of these EIPs are highly variable. Two parks are not places but are 'virtual EIPs' with materials exchanged on a regional network basis. The actual parks range in size from 3.5 acres to 7,000 acres. The physical settings also vary – six of the parks involve brownfield reclamation, three are in greenfield locations. While most EIPs were based on manufacturing, two focus on agricultural products, one includes marine technology and aquaculture and one involves sustainable harvesting of a second growth coastal-forested area. Three parks plan to provide scenic landscape or other recreational use in addition to economic use. Managing entities for the EIPs encompass government and private sectors, including cities, counties, towns or their development authorities, local economic development corporations, private industry and other non-government organisations. The parks are in various stages of development, with a few still in the design stage, some with baseline studies underway or completed, and several already in the recruitment phase. No parks are fully implemented as yet.

Such eco-parks aim to orient companies to look at wastes as commodities to be sold or recycled with the aim of boosting companies' financial performance, while reducing the demand for virgin feed stock materials, minimising the cost of treatment and disposal of wastes and alleviating the adverse environmental impact of industrial development. The most developed examples, such as Kalundborg, evolved over twenty-five years, yet there is some potential for policy intervention to begin such developments. Key investments can be made in assembling sites and providing supporting infrastructure, as well as identifying and approaching potential partner companies. The system needs to be simple enough for the participating companies to work with, yet not allow the demise of one company to bring down the entire project.

There are also variants of this eco-park theme being developed in the UK which seek to improve energy efficiency and encourage greater environmental responsibility among park occupants. Examples of these can be found at Swaffham in Norfolk, Stroud and Wales. The Ecotech Innovation and Business Park at Swaffham includes an education and training centre to provide support for business on-site, plus a visitor centre. The commercial units will be energy efficient and heated from a biomass-fired district heating system. A wind turbine guarantees the power supply with any excess sold off to the National Grid. The Park was intended to act as a catalyst for inward investment and received central government funding through Regional Challenge and Rural Challenge. As with most UK bids for government funds, the Park was developed by a partnership comprising Norfolk County Council, Norfolk and Waveney Training and Enterprise Council, City College Norwich, Easton College, University of East Anglia, Rural Development Commission, Fyfield Estates, Real Architecture, City and County Developments Ltd and LRZ Bio Energy Systems. Two companies indicated that they will move to this site – both have environmental themes – Environmental Vehicle Systems and Advanced Natural Fuels. Phase I of the Innovation and Business Park involved the construction of 2,465 square metres of 'eco-tech units' and 4,200 square metres of business units. These were expected to create 120 full-time and 80 part-time jobs. The associated tourist attraction takes up 2 hectares of the site, including a 4,500-square metre visitor centre, including conference and seminar facilities. The total value of the project is estimated at £12m with the potential to generate 500 jobs.

At Ebley, near Stroud, the Energy 21 Renewable Energy Park is being built to act as a showcase for renewable energy. The park will consist of a range of renewable energy equipment, including water turbines and solar energy. Research, education and employment opportunities in renewable energy generation and use will be provided in an exhibition and training centre. The sale of electricity from the site will help to fund staff and running costs. The Dyfi Ecopark at Machynlleth in Wales uses methane from the local sewage plant for heating and occupants will have to conform to strict environmental conditions. Units on the site are built from local semi-durable softwoods and other locally sourced natural materials, with the aim of minimising environmental impact and enhancing the working environment at no greater cost than more conventional schemes (*Architects Journal*, 1997).

However, there are also a number of barriers to industrial ecology approaches (Brand and de Bruijn, 1999). These can be subdivided into:

- Technical barriers: including the possibility that local industries do not have the potential to 'fit together';
- Informational barriers: which make it difficult to find new uses for waste products, relating to poor information regarding potential markets and potential supplies;
- Economic barriers: the incentive to use waste streams as a resource depends in part upon there being a reliable market;
- Regulatory barriers: current regulatory structures may prevent industries or industrial processes being linked together;
- Motivational barriers: firms, public sector agencies and other relevant local actors must be willing to co-operate and commit themselves to the process. Trust is a key factor here and companies may be unwilling to provide information about production processes and (by)products for competitive reasons.

A more fundamental criticism is that industrial ecology may make the partners in such projects overdependent on each other. Creating connections between companies may effectively 'fix' infrastructure and lessen the chances of firms' movement. For a locality, this may be a benefit, but it could also be argued that such inflexibility might lead to longer term inertia and a lack of innovation within companies.

Indicators and output measurement

The projects and initiatives outlined in this chapter need to be developed in tandem with a set of indicators to measure, at the very least, some form of progress towards environmental improvement. There is a growing body of such indicator work to draw upon and:

> This work should make it easier to 'tell the difference' between a plan which has real potential for environmental leverage on economic development and one which does not; and between one which moves beyond a balancing conception on sustainable development to one based on a relational approach to environmental capacities.
>
> (Healey and Shaw, 1994: 435)

In Cumbria, for example, the county council conducted a sustainability appraisal of economic development using a list of criteria under three main headings against which economic development strategy could be assessed (Table 5.2).

There is a wealth of research work on indicators to draw upon which needs to be integrated with these initiatives to measure progress. Such work includes the UK Local Government Management Board's Sustainability Indicators Research Project and that of the New Economics Foundation, as well as well-publicised individual examples such as those in Lancashire and Seattle. In areas eligible for EU Structural Funds, the development of such indicators and assessment procedures is likely to be essential in order to assess ERDF-funded projects in terms of their sustainability outputs. This reflects recent European Commission concerns about the failure to meet environmental requirements within the Structural Fund programme at the regional level (Keller, 1997), resulting in proposals to 'green' the Structural Funds by developing a set of environmental indicators to assess the impact of projects and strategies (European Commission, 1997). Indeed, a joint seminar held by the European Commission and the Environment Agency sought to encourage the use of Structural Funds to implement 'environmentally sustainable development' in eligible regions (European Commission/ Environment Agency, 1997). This recommended developing environmental baselines against which to measure performance, integrating environmental profiles into Structural Fund programmes and encouraging adoption and diffusion of clean technologies and eco-industries. Some pioneering work in

Table 5.2 Criteria for sustainability appraisal of economic development in Cumbria

Quality of life and local environment	Natural resources	Global sustainability
• Community participation	• Air quality	• Climate change
• Access to goods/services	• Land	• Biodiversity
• Personal safety	• Soil quality	• Deforestation
• Basic needs	• Mineral stocks	
• Education/training	• Freshwater	
• Employment	• Sea	
• Wealth creation		
• Health		
• Neighbourhood cohesion		
• Landscape quality/character		
• Built environment		
• Resource use		

Source: Adapted from Forum for the Future (1998b).

relation to this has been undertaken by North East Lincolnshire Council (1998) in the UK.

Conclusions

This chapter has outlined the types of activity that have been developed in response to the sustainability agenda and which combine, in varying degrees, economic, social and environmental objectives. From this analysis, we can draw a number of conclusions and make some suggestions as to how the concept of sustainable local and regional economies can be progressed and implemented. One of the key features which emerges is that there has certainly been a growth in such initiatives since the late 1990s. Although it is sometimes difficult to sort out the plethora of individual initiatives and schemes, there would appear to be no shortage of good examples and ideas for practitioners to draw upon in devising their own initiatives. In policy terms, taking these good ideas and initiatives forward means locating such sustainable economic regeneration developments within broader frameworks of purposive policy action at the local and regional scales, as elements of a broader shift towards an economy based upon sustainable development. As such then, these local initiatives on their own may be piecemeal and have very little impact upon the wider economies in which they are placed. Their importance lies in the sense that they indicate what I have elsewhere termed 'waystations' (Gibbs, 1993) or exemplars of what can be done. This need for locating initiatives within broader frameworks would appear to be a vital element in future developments. For example, in the UK context the revived regional agenda through Regional Planning Guidance and the Regional Development Agencies has given rise to regional economic strategies and planning guidelines. At present it is not clear whether sustainability will be taken seriously within such developments. In the UK and elsewhere there is a clear need to see how local strategies for sustainable economic regeneration should fit in with broader regional and national policies. Moreover, for policy makers there is also a need to think through how such integrative economic and environmental strategies can be integrated with other local level policies, which also have a bearing on economic, environmental and social concerns. For example, the UK Environment Agency has a set of Local Environmental Action Plans (LEAPs) based on catchment areas which although developed in consultation with industry, local authorities and pressure groups, are narrowly technocratic watershed management plans which do not connect with any substantive economic or social agendas.

Moreover, while there is much material available on individual initiatives, without in-depth analysis it is difficult from to assess their relative importance *vis-à-vis* other initiatives and agendas within their local areas. Thus it is difficult to assess whether they form a central part of regeneration strategies or whether they are marginal. In addition, it is impossible to assess the impact of these initiatives as few would appear to have readily identifiable measures of progress. In conjunction with the need for a more holistic regional and local strategy, any strategic development needs to encompass indicators and output measures for sustainable economic regeneration. While there has been a great deal of research on the theme of indicators, this has rarely been taken up as a means to measure progress on economic regeneration, reinforcing a view that sustainability is a marginal concern to the 'real business' of economic regeneration measured through GDP, jobs created, firms assisted, etc. For example, the UK government's own indicators proposed for use by the RDAs are symptomatic of this – the only indicator proposed for sustainability is the percentage of new houses built on previously developed land as a core indicator and net hectares of derelict land brought into use as a possible additional sustainable development indicator. For policy makers, developing a set of core indicators and measures acceptable across local areas and regions might also provide a useful lobbying tool to engender a shift in national governments' own regeneration output measures. This is not to say that this is a particularly easy task, as the debate over which indicators to use and what they indicate reveals (see the special issue of the journal *Local Environment* 4(2), 1999 for a useful collection of articles on sustainability indicators and measures). Combined with the European Commission's work on incorporating environmental measures into the Structural Funds, mainstreaming environmental and social measures into economic regeneration may be one way of getting the issues taken seriously and not relegated to token status.

One interesting development is the way in which some rethinking has been underway concerning more conventional economic regeneration strategies. One area examined in this chapter is that of inward investment, where there has been a move to incorporate environmental and social aims into more conventional inward investment policies. As a caveat it is worth pointing out that although inward investment has formed a key component of many economic regeneration strategies, success in attracting any conventional investment, let alone that related to the environment, has proved elusive in many peripheral areas.[1] Indeed, some areas have placed an increasing emphasis upon the development of indigenous enterprise as opposed to inward investment strategies. There is potential to encourage small firm advisory bodies to provide advice to small firms on sustainable development. Given the difficulties of reaching such firms and, in some cases, of actually having an influence upon firm behaviour, this may need to be linked to other initiatives such as local public sector purchasing policies and supply chain networks. While encouraging supply chain links appears to be an area with potential, there are issues here regarding the extent to which economies can be (or are) self-contained. The role of employment initiatives linked to environmental and social aims is also worthy of attention. Of interest here is the potential to create jobs through environmental initiatives where there has been much research at a national level, but relatively little work on the local and regional consequences. In addition there is scope for more detailed work on intermediate labour market schemes. In the UK some attempt at this has been made through the Environmental Task Force component of the Labour government's 'New Deal' initiative, which attempts to provide training for unemployed people.

Another innovative development has been in infrastructural development, notably at the level of individual buildings, but also through the development of clusters of office buildings and industrial units. Such developments are frequently very small scale at present, but there would seem to be scope for investigating the potential for encouraging these in future with the aim of making such developments mainstream. For 'new build' parks this may also be combined with layouts and buildings that encourage energy efficiency and utilise renewable energy sources. Obviously much may depend upon cost compared to more conventional construction and the savings accrued through energy savings. Related to this is the growing interest in eco-industrial parks where companies are linked through the exchange of wastes and the provision of common environmental services. The difficulty of establishing such parks may be illustrated by continual reference in the literature to the example of Kalundborg as the only operating park and the early development stages in the USA. However, there may be some interesting lessons to be learned from these US examples and this is an area which is receiving increasing attention within the US and Europe. For some localities there may also be potential to encourage the development of high technology sectors that can contribute to sustainability – the growing global market for environmental technologies and services provides an opportunity here. Many of the firms in these sectors are small and there may be opportunities for networking such firms to assist with global market access, as well as exchanging information and joint project bids. There is thus a role for policy makers to act in a proactive manner to create the kinds of networking and 'institutional thickness' associated with cluster development.

Whatever initiatives and strategies are developed, a key issue is that different localities will be starting from different points in terms of their environmental, economic and social inheritances – there will therefore be no 'one best way' for localities to proceed, nor a detailed blueprint that they can follow. To reiterate, there also needs to be a strong driver for change emanating from national governments. Frequently, national regeneration funding regimes do not encourage the development of sustainable economic regeneration, neither in terms of many of the output measures used, nor in terms of their frequently short time-scales. Changes are needed to the major funding regimes, such as the Single Regeneration Budget in the UK and the European Regional Development Funds in the EU, to encourage local government and other partners to respond to the sustainability agenda. A shift in thinking in many localities may only come when it is perceived that sustainability is a major concern of central government and that regeneration strategies must address the issues, rather than sustainability being marginal to more mainstream initiatives. However, how far local and regional economies can shift to a more sustainable basis remains open to contention. This then leads us back to some of the questions raised at the outset of this book about the links between theory and practice and these are reconsidered in the next, and final, chapter.

6

INTEGRATING ECONOMIC DEVELOPMENT AND THE ENVIRONMENT
Future prospects for local areas

Introduction

An important rationale for this book has been the need to explore in some depth the potential to integrate economic development and environmental policies at the level of local areas. It was initially argued that much of both the academic and policy literature views 'the local' as a key site where positive action can be taken to implement such integrative policies. Debate over the most effective course to take, problems of implementation and the contested nature of sustainability remain important of course, but there is a substantial consensus of opinion that local scale policy and measures are of key importance. While sympathetic to this argument, it was felt that this position needed to be justified and subjected to critical examination, rather than simply taken for granted. To this end, while the focus of the book has similarly been upon 'the local', it has also been explicitly argued that local level actions are not enough in themselves and that they must be contextualised both nationally and internationally.

Previous chapters have therefore explored both the economic and environmental policy contexts for local sustainable development at wider scales than just the local. One argument has been that any interpretation of the potential for local action needs to be firmly placed within a theoretical context – something that is frequently lacking in many accounts. This is not just on the grounds that such an approach enables a better understanding of the processes at work (although it obviously does), but also on the grounds that better conceptualisation facilitates an understanding of the potential for, and limits to, practical policy development (Gandy, 1997). As Roberts and Gouldson (2000: 258) comment 'there is . . . a clear need for the development of over-arching theory and for policy guidance that can help to explain and predict the nature and outcomes of the relationship between economic development and environmental management'. In this chapter I revisit some of the theoretical arguments contained in Chapter 1, as well as reflecting on the limits to local approaches in a world increasingly dominated by the activities of large corporations and the drive towards free markets and trade. Finally, I outline the issues that need to be considered in the future as a way forward for local economies and consider some of the strengths and weaknesses of a local approach.

Reflections on theory

Chapter 1 contained an examination of the ways in which authors have approached the integration of economic development and the environment. For those writing from a specifically environmental perspective a major focus has been upon the two approaches of sustainable development and ecological modernisation. While these have their merits as a means of explanation, it was argued that there is considerable advantage in combining these with ideas from political economy approaches, in particular urban regime and regulation theories. While those writing within the latter two approaches rarely address environmental issues, drawing upon such political economy perspectives helps to illustrate the key importance of the role of the state and the need for interventionist approaches at all spatial scales. It is therefore not sufficient to develop local policy and initiatives in isolation, but we need to see these as an entry point into the intersecting processes operating at multiple spatial scales.

Indeed, part of the 'capacity to act', in urban regime terms, and the 'situative context', in ecological modernisation terms, within local areas is comprised of the opportunities and constraints that are imposed by forces acting at broader spatial scales. It is perhaps only at these levels that rules can be institutionalised and equipped with sanctions to assist in moving towards new modes of regulation. Having said this though, it is also clear that the task of engendering a shift at local, national and international scales simultaneously is a massive one and thus we return to the need to see local developments as a form of experimentation and exploration for higher spatial scales. Some authors would argue that this in itself is problematic. For example, Hay (1994: 223, emphasis in original) argues that 'such examples of *apparent* localised (and occasionally nationalised) environmentally sustainable economic growth are *parasitic* upon the global capitalist growth imperative, which is itself the very antithesis of environmentally sustainable economic growth'. Indeed, several authors have strongly criticised the concepts of ecological modernisation which propose that an 'ecologically sustainable capitalism' can be developed for this very reason (see, for example, Goldblatt, 1996; O'Connor, 1996). Mol and Spaargaren sum up (whilst not agreeing with) such views as:

> Neglecting capitalism and failing to attack the fundamentals of the real capitalist world order will result in superficial and cosmetic environmental reforms that are unable to resolve the ecological

crisis in any fundamental way. Moreover, such measures will rather strengthen the capitalist mode of production as it makes capitalism less in need of a green critique and it promotes and facilitates the continuation of established socio-economic practices to the benefit of those in power.

(Mol and Spaargaren, 2000: 22)

While sympathetic to those critiques of ecological modernisation which highlight the excessive technological optimism of some authors, the argument that is being proposed in this book is for a form of incrementalism as perhaps the best that can be achieved in current circumstances. I recognise that there are longstanding debates over the merits of radical ecologism as against moderate environmentalism (reflecting the strong and weak sustainability arguments in Chapter 1) but there has not been sufficient space here to explore these in detail. While not wishing to underestimate the scale of the shift needed, drawing upon some of the theoretical ideas outlined in Chapter 1 does provide some grounds for optimism in the medium to longer-term. In their different ways, both ecological modernisation and regulation theory open up the possibility of change. In the case of the former, ecological modernisation approaches argue that capitalist economies can be shifted onto a different development path, one which encompasses continued economic prosperity and environmental protection, assisted by the application of clean technologies. From a much broader perspective on economic and social development, regulationist approaches alert us to the possibility that such new forms can be developed and that there are no necessary outcomes. Rather, the outcomes for economies and societies are open to debate and shaping. Indeed, the accumulation system may be able to incorporate environmental issues as new arenas for growth, although whether this always represents a positive outcome in environmental or economic terms is also open to debate (Bridge and McManus, 2000).

Attempts to integrate ecological modernisation and regulationist approaches may therefore yield greater explanatory power on the potential for integrative policy attempts. Recent work from a regulationist perspective has explored how certain fractions of capital have appropriated ecological modernisation as a means of legitimating their activities, albeit that this has largely been a cover for 'business-as-usual' by the transnational corporations concerned (Bridge and McManus, 2000). Moreover, recent work on ecological modernisation has begun to recognise the need to incorporate a social context, arguing that environmentally sound production and consumption is a possibility with changes to the mode of production and under different relations of production (see, for example, Mol and Spaargaren, 2000). In regulationist terms, such 'new modes of social regulation' may be possible, involving not only the 'real regulation' of laws and controls, but also changes in values and attitudes supportive of new modes of production and consumption. The accumulation system may thus be able to incorporate environmentalism as a new arena for growth (Bridge and McManus, 2000). More fundamentally, it presumes a regime of accumulation which is different to the current one (Walker et al., 2000).

However, while a regulation approach may provide a useful framework within which we can analyse how these changes might occur, it is frequently difficult to see how any substantive shift will occur in a world largely dominated by potential consensus on free markets and free trade as the basis of development. As the examples of Australia and Japan in Chapter 3 revealed, even where there has been a shift towards sustainable development as a guiding principle, this can soon be abandoned in the face of short-term economic difficulties. Similarly, the lack of engagement with the sustainability agenda by the USA, simultaneously the world's largest economy and largest source of carbon emissions, does not give grounds for optimism. In the short period of time between being sworn in as President of the United States and the time of writing, George W. Bush and his administration have abandoned the Kyoto Treaty on carbon emissions, dropped controls on arsenic levels in drinking water, indicated approval for oil exploration in sensitive environmental areas in Alaska and Florida and proposed the construction of large numbers of nuclear power stations. For local areas in the USA then, the task of introducing policies to integrate economic development and the environment is made that much more difficult by the lack of a supportive 'situative context' nationally and the 'objective limitations on successful intervention' (Jänicke, 1997: 4) this imposes. In some cases then, as Gandy indicates:

> If we extend our conception of Fordist crisis to embody a mismatch between economy, society and nature, then it is clear that, far from solving the contradictions of Fordism, the deregulative momentum of the post-Fordist era is exacerbating the ecological contradictions of capital.
>
> (Gandy, 1997: 351–2)

To counter this, albeit perhaps not at the same level of magnitude, one can point to the examples of both Sweden and the European Union as a whole as examples of more positive moves towards a policy of ecological modernisation and attempts to effectively *re*-regulate the relationship between economy and environment. Certainly the Bush administration's stance on the Kyoto Treaty has been followed by political calls for the EU to take a global leadership role on carbon emission reductions, as well as the more pragmatic argument that this provides European industry with an opportunity to take a lead in renewable energy technologies and achieve competitive advantage over US industry. A key point then is that sustainable development is inherently a political concept open to debate and contestation. It involves political struggles over power, given that it represents a challenge to the established economic and social order. Even ecological modernisation, while working with the grain of capitalism, entails a major shift in attitudes and practices. As Healey and Shaw indicate:

> The challenge for the new environmental agenda is therefore not simply one of developing appropriate conceptions, policy instruments and skills in local operationalisation. It is a political challenge for real leverage over economic discourse – at the level of policy and practice. Only if this happens will the sustainability objective of a beneficent relation between economic development and global/local environmental quality be delivered.
> (Healey and Shaw, 1994: 434)

This is important, given that much ecological modernisation research has relatively little to say about barriers to implementing ecological modernisation, other than seeing it as state failure, whereby 'policy becomes *locked-in* to a reactive and standardised approach even though more proactive policies are available and might offer economic and environmental advantages' (Gouldson and Murphy, 1997: 80, emphasis in original). The role of the state, particularly in some of the early ecological modernisation accounts is seen as minimal (see, for example, Huber, 1982, 1985) and in others as performing an enabling function. To date, ecological modernisation has had relatively little to say about the form of institutional adaptation of change needed at nation-state level, let alone at the local scale. The assumption is that some form of enabling state will deliver ecological modernisation through corporatist relationships between government and industry, although co-opting environmental movements where necessary, thus ignoring issues of participation and reducing the rest of society to passive consumers to be provided with enough information to make informed (but market-based) choices.

Weaker versions of ecological modernisation thus largely ignore the major institutional changes needed, despite being based on the notion that the necessary changes to institutions *can* be made (Christoff, 1996). One criticism from a theoretical approach is that, until recently, ecological modernisation has frequently failed to take account of the social processes at work, such that it has relied upon a narrow technocratic and instrumental approach rather than being integrative and communicative. This approach will not lead to the type of embedded cultural transformations that will sustain factors such as environmental improvements, reduced consumption and greater equity (Cohen, 1997, 1998; Jamison and Baark, 1999). If ecological modernisation is to continue to offer any useful theoretical or practical guidelines then, as suggested in Chapter 1, a fruitful avenue to explore is Jänicke's (1992, 1997) notions of the capacity for action through investigations of strategic capacities, structural frameworks and the role of actors, which has parallels with urban regime approaches. This would assist in moving away from the idea that implementing policy and devolving it to lower spatial scales is a largely problem-free experience. Notions of the capacity for action highlight the processes of experimentation, struggle and conflict in environmental policy as opposed to its objective promotion as ecological modernisation.

Given that the problems are largely political, then a focus upon the types of governance structures conducive to sustainable development is needed and this again points to the merits of drawing upon concepts from political economy as a means to help us understand the prospects for change. It is argued that drawing upon urban regime theory and regulation theory is useful in this context because:

> Despite considerable differences in their understanding of agency, attentiveness to extra-local influences, and ontological precepts regarding causality, urban regime theory and regulation theory nonetheless share a common concern with the emergence, maintenance, and dissolution of systems of governance.
> (Bridge and McManus, 2000: 14)

The approach adopted here has been to draw upon theoretical strands from work in both political economy and environmentalist approaches. This seems likely to yield greater explanatory power than any search for some form of single theoretical explanation, given the complex and dynamic nature of the relationship between the economy and the environment (Roberts and Gouldson, 2000). That this is an approach which is gaining ground is shown in the recognition by those working from an ecological modernisation perspective that congruence between environmental and economic aims is not pre-given but 'constantly (re)produced by struggles and clashes between diverging interests, changing ideologies, and historical transformations in other social arenas' (Mol and Spaargaren, 2000: 24). As Buttel (2000: 58) argues, 'a full-blown theory of ecological modernisation must ultimately be a theory of politics and that state – that is a theory of the changes in the state and political practices . . . which tend to give rise to private eco-efficiencies and overall environmental reforms'. Combining ecological modernisation with political economy approaches may therefore give rise to the type of comprehensive theoretical framework that Buttel envisages.

The strengths and weaknesses of a local approach

One advantage of focusing upon the local scale is that it does allow the development of demonstration policies and initiatives that could be introduced at wider spatial scales. This does not mean that such local sustainability means self-containment or isolation. Rather it involves 'the development of local-global relationships conducive to sustainability' (National Science Foundation, 2000: 8). In addition we need to bear in mind the dual nature of localities, as both places and sites of everyday life and as nodes in a global network of flows and interactions. While policy may be focused on the former, the interdependent nature of places must be considered. Local areas cannot become more sustainable by making other places less sustainable. The types of initiatives outlined in Chapter 5 need to be seen as a means of experimentation with the most appropriate actions to take. They are less 'model' practices to be replicated elsewhere, but rather are one part of the identification of the processes and conditions through which sustainability is more, or less, likely to be achieved (National Science Foundation, 2000). The progress of the sustainability agenda through local level action will come about not just through this local level activity itself but also, at the same time, through pressure on wider spatial scales of governance, reflecting this role of local places as nodes in global networks and flows. As Low *et al.* argue, the capacity:

> to develop policies supportive of ecological sustainability is dependent both on the local progress of the discursive trope and on the framework provided by national leadership, financial support, and market regulation. But city governments can also exert upward pressure on the nation state to engage with the ESD [Ecologically Sustainable Development] agenda and provide working examples of ESD practices.
>
> (Low *et al.*, 2000: 292)

Overall then we should not discount the possibility of change, even if the dominant economic discourse remains one of free trade, competitiveness and primacy of markets to the relative exclusion of environmental and social issues. Recent protest over the legitimacy of this discourse, in the form of street protests at G7 and WTO meetings, acts as a challenge that may yield change towards a greater concern with the environmental and social implications of global economic development. In particular, the discourse of ecological modernisation may constitute a powerful mechanism for change. It can be criticised for legitimating existing economic activities, albeit with a 'green tinge'. However, such 'environmental discourses represent a significant component of a mode of regulation that may be emerging to stabilise contradictions inherent to the relationship between . . . capital and environment' (Bridge and McManus, 2000: 20). In the longer term it may act as an important catalyst to shift both political and public opinion. As such, ecological modernisation may be able to act as a 'Trojan horse' to encourage a movement from weak to strong forms of sustainability, with rather more chance of success than calls for a wholesale restructuring of economic and social life. This transformation will in part be engendered by changes in governance and particularly much greater democratic political regulation of market systems. Action at the local level can only go so far in fostering these changes. As Low *et al.* state in relation to cities:

> Action at the level of the city will hardly be sufficient. City governments need discursive, regulatory and financial support from their national and regional governments. They must demand it. Nation-states are still the principal actors at global level and their interactions will shape the regulatory regimes which must eventually limit the competitive freedom of both cities and corporations.
> (Low *et al.*, 2000: 303)

Local action therefore needs international and national framework agreements upon which to base local policy. While the EU's Environmental Action Programmes and, to a lesser extent, Agenda 21 provide this framework, there is still a need to develop appropriate forms of organisational and operational capacity to integrate policy. This should be developed at the most appropriate spatial scale, whether local, national or international. Meeting the challenge posed by sustainability requires:

- Integrated approaches within strategic frameworks which allow policy instruments to address multiple problems simultaneously;
- Policy intervention at the local level, as opposed to passing on the problem either spatially or temporally;
- Policy solutions which lead to changes in individual patterns of consumption and behaviour by individuals and companies (European Commission, 1998a).

The development of such forms will be through a process of negotiation by a wider set of actors than simply government alone (Roberts and Gouldson, 2000). As the example of Sweden indicates, involving a wide and varied coalition of actors and building public support is a key part of the process.

While previous chapters have outlined what can be achieved at the local scale, it remains important to bear in mind the limits to such initiatives in the context of broader national and international economic developments. There are a number of constraints upon taking action at the local scale. First, assuming a capitalist economy, it may well be difficult to encourage companies to move towards sustainability at a faster rate than markets allow without losing their competitive advantage. Such market constraints include (EU Expert Group on the Urban Environment, 1994):

- The present low costs of energy, materials and waste disposal compared to other factors of production. This is especially true compared to the cost of labour, where labour shedding may occur at the expense of higher resource impacts;
- The lack of signals provided by the market allowing companies to distinguish between sustainable and unsustainable patterns of resource use;
- High discount rates and rates of return on investments discourage investment in resource efficiency measures other than those with very short payback times.

Thus any local or regional policies which try to force companies to go further than existing 'best practice' is likely to be counterproductive. This will only be remedied if action is taken at national and international levels such as ecological tax reform, developing national environmental business strategies, promoting longer term investment patterns and creating markets for sustainable businesses. Specifically local action: 'is inevitably constrained by the limitations on the local authority's influence on private business and individual behaviour, and by the limits placed by central government on local authorities' ability to restrain 'unsustainable' activities and decisions in the private sector' (Thomas, 1994: 154).

Conversely, there are arguments that in future companies will need to take much greater account of the 'triple bottom line' of economy, environment and equity if they are to remain competitive. Even within existing capitalist economies, there are likely to be important competitive advantages to be gained from a greater engagement with environmental concerns, not to mention the benefits of first-mover advantage in areas such as fuel cell technology and renewable energy (von Weizsäcker et al., 1997). Even so, while local areas and regions can promote themselves as offering a high quality environment and economy based on environmentally friendly economic activities, then the total amount of such activity that can be supported is limited by broader economic factors. For example, demand for recycled products may be limited in which case any expansion by one area in this field simply occurs at the expense of other areas to give a zero sum gain overall. As with more conventional local and regional economic strategies that attempt to recreate Silicon Valley in peripheral areas, only a limited number of local areas and regions are likely to become leaders in sectors such as environmental technologies.

Another major hurdle is the difficulty of local strategies developing around an environmental focus in the absence of a wider complementary economic framework. The task for governance structures then is to ensure that governing powers are sutured together in a well-integrated system. However:

> A practical difficulty with implementing an environmental agenda relates to what might be called the jurisdictional ambivalence of environmentalism. This ambivalence is reflected vertically (in terms of levels of government and administrative systems) in the concern of environmentalists for regional, national, and even international solutions to many problems, at the same time as the most active focus of much environmental political activity is the local community or even the household. The ambivalence is also reflected horizontally, as environmental concerns overlap with and cut across so many traditional structures of contemporary political and institutional organisations ... one result of this vertical and horizontal ambivalence is that traditional administrative and policy solutions to the environmental challenge are difficult to design.
>
> (Stren, 1992: 313)

While the theoretical analysis opens up the practical possibilities of change and the notion of 'sites of intervention' at various spatial scales to achieve this, there remain formidable problems to this occurring. Chapter 2 indicated some of the issues here revolving around the forms of capitalism that have developed based upon free trade and the dominance of transnational companies. The dominant economy remains the USA where the federal government remains sceptical about the existence of global warming, let alone the merits of sustainability as a policy direction. The Bush administration's rejection of the Kyoto Treaty to reduce carbon emissions was based, in part, on the assertion that implementing the Treaty would harm US competitiveness. As Low *et al.* (2000: 287) comment 'in America it is benefits to America which have to be stressed'. Indeed, Chapter 3 indicated that differing national environmental policy contexts have been an important factor in determining what is possible at the local level. In the USA:

> the vertical networks around environmental issues are very weak. State legislatures in particular show little willingness either to encourage local government or to formulate Agenda 21 policies of their own. There is no connecting tissue between tiers of government in environmental matters.
>
> (Low *et al.*, 2000: 288)

Similarly, Chapter 3 indicated that the national context can go into reversal as the case of Australia indicates, where there has been a move away from a firm commitment to what was termed 'ecologically sustainable development' to a much more superficial interpretation. As with the USA, the argument has been that a strong line on sustainability will be detrimental to the country's economic interests and competitiveness at a time of growing economic globalisation. In summary then:

> The underlying rationality of democratic representation encourages states to restrict their responses to such crises to the minimum they perceive necessary for short-term restoration of legitimacy. This is likely to be achieved through a combination of symptom amelioration, token gesturism, the 'greening' of legitimating political ideology, and the displacement of the crisis in a variety of different dimensions: either downwards into civil society (by making individuals responsible on a personal level for the response to environmental crisis through facilitation of 'green' consumer choice); or upward onto a global political agenda; or indeed, sideways in presenting the crisis as another body's (e.g. state's) legitimation problem.
>
> (Hay, 1994: 221)

Conversely, Sweden remains perhaps one of the few examples to indicate how 'economic prosperity, a modern lifestyle, social and environmental justice and ecological sustainability are far from mutually exclusive' (Low et al., 2000: 297). One of the strengths of the Swedish case is the involvement of both government and NGOs in constructing a broader national consensus through public education that economic development should be focused upon ecological modernisation lines. Even here though, it is important not to get too enthusiastic about Sweden as a potential role model. Opposition to the policy has come from some sections of Swedish industry and while ambitions may be high, the realities of economic life mean that Sweden has not reduced carbon emissions and retains a very large ecological footprint impacting elsewhere in the world.

In total then it is hoped that this book helps to put the arguments about integrating economic development and the environment into sharper focus. Past research work in this area has frequently focused almost exclusively upon the themes contained within Chapters 4 and 5. In particular, there has either been a focus upon the broad themes and issues that need to be considered, as outlined in Chapter 4, without being too specific about concrete policy, or, as in Chapter 5, outlining the large number of initiatives and policies in place which attempt to implement aspects of sustainability, without seeing these within any broader strategic context. In particular the argument here has been that while these are worthy efforts in themselves, without a clearer understanding of sustainability and the environment as a contested and political issue (or set of issues), then little headway will be made with policy implementation at any scale. Finally, it will be obvious that the focus of this book has been solely upon the developed world. Lack of time, space and expertise have not allowed a consideration of the situation in the developing countries where the bulk of the world's population lives. However, given the disproportionate contribution to environmental problems from the developed world, this focus seems justified. Moreover, in the absence of action by developed states it seems hypocritical to expect developing countries to engage with the sustainability agenda.

REFERENCES

Agnew, J. and Corbridge, S. (1995) *Mastering Space*, London: Routledge.
Altvater, E. (1993) *The Future of the Market: An Essay on the Regulation of Money and Nature After the Collapse of 'Actually Existing Socialism'*, London: Verso.
Amin, A. (1992) 'Big firms versus the regions in the Single European Market' in M. Dunford and G. Kafkalas (eds) *Cities and Regions in the New Europe: The Global-Local Interplay and Spatial Development Strategies*, London: Belhaven, 127–49.
Amin, A. and Malmberg, A. (1992) 'Competing structural and institutional influences on the geography of production in Europe', *Environment and Planning A*, 24, 401–16.
Amin, A. and Robins, K. (1990) 'The re-emergence of regional economies? The mythical geography of flexible specialisation', *Environment and Planning D: Society and Space*, 8(1), 7–34.
Amin A. and Thrift N. (1995) 'Institutional issues for the European regions: from markets and plans to socioeconomics and powers of association', *Economy and Society*, 24, 41–66.
Andersen, M. (1994) *Governance by Green Taxes: Making Pollution Prevention Pay*, Manchester: Manchester University Press.
—— (1997) 'Ecological modernisation capacity: finding patterns in the mosaic of case studies' in J. van der Straaten and S. Young (eds) *Ecological Modernisation*, London: Routledge.
Andruss, V. C., Plant, J. and Wright, E. (1990) *Home! A Bioregional Reader*, Gabriola Island, British Columbia and Philadelphia: New Society Publishers.
Angel, D., Jonas, A. and Theyel, G. (1995) 'Constructing consensus: environmental policy making in Massachusetts', *Urban Geography*, 16(3), 397–413.
Architects' Journal (1997) 'Green light for businesses', *Architects' Journal*, 20 February, 31–5.
Artibise, A. (1995) 'Achieving sustainability in Cascadia: an emerging model of urban growth in the Vancouver-Seattle-Portland corridor' in P. K. Kresl and G. Gappert (eds) *North American Cities and the Global Economy: Challenges and Opportunities*, Thousand Oaks, CA: Sage, 221–50.
Bakshi, P., Goodwin, M., Painter, J. and Southern, A. (1995) 'Gender, race and class in the local welfare state: moving beyond regulation theory in analysing the transition from Fordism', *Environment and Planning A*, 27, 1539–54.

REFERENCES

Barrett, F. D. and Therival, R. (1991) *Environmental Policy and Impact Assessment in Japan*, London: Routledge.

Bassett, K. (1996) 'Partnerships, business elites and urban politics: new forms of governance in an English city?', *Urban Studies*, 33(3), 539–55.

Benton, L. M. and Short, J. R. (1999) *Environmental Discourse and Practice*, Oxford: Blackwell.

Bianconi, P., Gilioli, M. and Marinelli, M. (1998) 'STENUM Project: Bologna Sustainable Production Programme', paper presented to the Seventh International Conference of the Greening of Industry Network, Partnership and Leadership – Building Alliances for a Sustainable Future, Rome.

Blowers, A. (1993) *Planning for a Sustainable Future*, London: Earthscan.

—— (1994) 'Environmental policy: the quest for sustainable development' in R. Paddison, J. Money and W. Lever (eds) *International Perspectives in Urban Studies*, Volume 2, London: Jessica Kingsley, 168–92.

—— (1997) 'Environmental policy: ecological modernisation or the risk society?', *Urban Studies*, 34(5-6), 845–71.

Boehmer-Christiansen, S. and Murphy, J. (1997) 'Ecological modernisation – whose capacity is being built? Conflicting evidence from the UK' in L. Mez and H. Weidner (eds) *Umweltpolitik und Staatversagen*, Berlin: Edition Sigma.

Bovaird, T. (1994) 'Managing urban economic development: learning to change or the marketing of failure?', *Urban Studies*, 31(4/5), 573–603.

Brand, E. and De Bruijn, T. (1999) 'Shared responsibility at the regional level: the building of sustainable industrial estates', *European Environment*, 9, 221-31.

Bridge, G. and McManus, P. (2000) 'Sticks and stones: environmental narratives and discursive regulation in the forestry and mining sectors', *Antipode*, 32(1), 10–47.

British Telecommunications (1992) *A Study of the Environmental Impact of Teleworking*, Report by British Telecommunications Research Laboratories, Martlesham.

Bristow, H., Cope, D. and James, P. (1997) *Information and Communication Technologies: Implications for Sustainable Development*, Cambridge: UK Centre for Economic and Environmental Development.

Brown, V. A., Orr, L. and Smith, D. I. (1992) *Acting Locally: Meeting the Environmental Information Needs of Local Government*, Canberra: Centre for Resource and Environmental Studies, Australian National University.

Bruff, G. and Wood, A. (1995) 'Sustainable development in English metropolitan district authorities: an investigation using Unitary Development Plans', *Sustainable Development*, 3, 9–19.

Bührs, T. and Aplin, G. (1999) 'Pathways towards sustainability: the Australian approach', *Journal of Environmental Planning and Management*, 42(3), 315–40.

Buttel, F. H. (2000) 'Ecological modernisation as social theory', *Geoforum*, 31, 57–65.

Carew-Reid, J., Prescott-Allen, R., Bass, S. and Dalal-Clayton, B. (1994) *Strategies for National Sustainable Development: A Handbook for their Planning and Implementation*, London: Earthscan.

Carter, N. and Darlow, A. (1996) 'LA21 – none of our business?' *Town and Country Planning*, 65(4), 103–5.

Casagrande, E. and Welford, R. (1997) 'The big brothers: transnational corpor-

REFERENCES

ations, trade organisations and multilateral financial institutions' in R. Welford (ed.) *Hijacking Environmentalism: Corporate Responses to Sustainable Development*, London: Earthscan, 137–56.

Castells, M. (1997) *The Power of Identity, The Information Age: Economy, Society and Culture*, Volume II, Oxford: Blackwell.

Christoff, P. (1996) 'Ecological modernisation, ecological modernities', *Environmental Politics*, 5(3), 476–500.

—— (1998) 'Degreening government in the garden state – environmental policy under the Kennett government', *Environment and Planning Law Journal*, 15(1), 10–36.

Christoff, P. and Low, N. (2000) 'Recent Australian urban policy and the environment: green or mean?' in N. Low, B. Gleeson, I. Elander and R. Lidskog (eds) *Consuming Cities*, London: Routledge, 241–64.

Churchill, D. and Worthington, R. (1995) 'The North American Free Trade Agreement and the environment: economic growth versus democratic politics' in F. Fischer and M. Black (eds) *Greening Environmental Policy: The Politics of a Sustainable Future*, London: Paul Chapman, 87–103.

Clarke, M. and Stewart, J. (1994) 'The local authority and the new community governance', *Regional Studies*, 28(2), 201–7.

Clarke, S. E. and Gaile, G. L. (1997) 'Local politics in a global era: thinking locally, acting globally', *Annals of the American Academy of Political and Social Science*, 551, May, 28–43.

Clement, K. and Bachtler, J. (2000) 'European Union perspectives on the integration of environmental protection and economic development' in A. Gouldson and P. Roberts (eds) *Integrating Environment and Economy: Strategies for Local and Regional Government*, London: Routledge, 25–38.

Cohen, M. (1997) 'Risk society and ecological modernisation: alternative visions for post-industial nations', *Futures*, 29(2), 105–19.

—— (1998) 'Science and the environment: assessing cultural capacity for ecological modernisation', *Public Understanding of Science*, 7, 149–67.

Cohen, S. (1995) 'Mackenzie Basin Impact Study: broadening the climate change debate', *Ecodecision*, Summer, 34–7.

Cohen-Rosenthal, E., McGalliard, T. and Bell, M. (1998) *Designing Eco-Industrial Parks: The North American Experience*, Ithaca, NY: Cornell University.

Commission of the European Communities (1990) *Employment in Europe*, COM(90) 290, Luxembourg: CEC.

—— (1991) *Europe 2000: Outlook for the Development of the Community's Territory*, COM(91) 452, Brussels: CEC.

—— (1992) *'Towards Sustainability' – A European Community Programme of Policy and Action in Relation to the Environment and Sustainable Development*, COM 92(23), Brussels: CEC.

—— (1993) *White Paper on Growth Competitiveness and Employment: The Challenges and Ways Forward into the 21st Century*, COM(93), 700 Final, Brussels: CEC.

—— (1994) *Europe 2000+ Co-operation for European Territorial Development*, Luxembourg: Office for Official Publications of the European Communities.

—— (1996) *Progress Report from the Commission on the Implementation of the European Community Programme of Policy and Action in Relation to the*

REFERENCES

Environment and Sustainable Development 'Towards Sustainability', COM (95) 624 Final, Brussels: CEC.

—— (2001) *On the Sixth Environment Action Programme of the European Community, 'Environment 2010: Our Future, Our Choice'*, COM (2001) 31 Final, Brussels: CEC.

Commonwealth of Australia (1992) *National Strategy for Ecologically Sustainable Development*, Canberra: Commonwealth of Australia.

Corson, W. H. (1994) 'Changing course: an outline of strategies for a sustainable future', *Futures*, 26(2), 206–23.

Coughlan, A. (1993) 'The green empire', *New Scientist*, 140(1893), 48–50.

Council for the Protection of Rural England (1994) *Greening the Regions*, London: CPRE.

Cox, K. R. (1997) 'Governance, urban regime analysis, and the politics of local economic development' in M. Lauria (ed.) *Reconstructing Urban Regime Theory*, Thousand Oaks, CA: Sage, 99–121.

Cox, K. R. and Mair, A. J. (1988) 'Locality and community in the politics of local economic development', *Annals of the Association of American Geographers*, 78, 307–25.

—— (1991) 'From localised social structures to localities as agents', *Environment and Planning A*, 23, 197–213.

Crabtree, T. (1990) 'Regenerating local economies', *Town and Country Planning*, June, 176–8.

Crowley, K. (1998) ''Glocalisation' and ecological modernity: challenges for local environmental governance in Australia', *Local Environment*, 3(1), 91–7.

Daly, H. E. and Cobb, J. B. (1989) *For the Common Good: Redirecting the Economy Toward Community: The Environment and a Sustainable Future*, Boston, MA: Beacon Press.

Dauncey, G. (1986) 'A new local economic order' in P. Ekins (ed.) *The Living Economy: A New Economics in the Making*, London: Routledge, 264–72.

de Rosario, L. (1992) 'Green at the edges: Japan lifts environment-related assistance', *Far Eastern Economic Review*, 155(10), 39.

Department of the Environment (1992) *Protecting Europe's Environment*, London: DoE.

Department of the Environment, Transport and the Regions (1997a) *Building Partnerships for Prosperity: Sustainable Growth, Competitiveness and Employment in the English Regions*, Cm 3814, London: HMSO.

—— (1997b) *Sustainable Local Communities for the 21st Century*, London: HMSO.

—— (1998) *Sustainable Development: Opportunities for Change, Consultation Paper on a Revised UK Strategy*, London: HMSO.

Department of Trade and Industry (2000) *DTI Sustainable Development Strategy*, London: Department of Trade and Industry.

Diffenderfer, M. and Birch, D. (1997) 'Bioregionalism: a comparative study of the Adirondacks and the Sierra Nevada', *Society and Natural Resources*, 10, 3–16.

Doughman, P. M. and DiMento, J. F. (2001) 'The nature of the NAFTA environmental side agreement: material and discursive effects on the ecological systems of North America attributed to the Commission for Environmental Co-

REFERENCES

operation', paper to 97th Association of American Geographers' Annual Meeting, New York.

Drummond, I. and Marsden, T. (1995) 'Regulating sustainable development', *Global Environmental Change*, 5, 51–63.

Drummond, I. and Marsden, T. (1998) *Conditions of Sustainability*, London: Routledge.

Dryzek, J. (1994) 'Ecology and discursive democracy: beyond liberal capitalism and the administrative state' in M. O'Connor (ed.) *Is Capitalism Sustainable? Political Economy and the Politics of Ecology*, New York: Guilford Press, 176–97.

—— (1997) *The Politics of the Earth: Environmental Discourses*, Oxford: Oxford University Press.

East of England Development Agency (1999a) *Towards a Regional Economic Development Strategy for the East of England: A Consultation Draft*, Cambridge: EEDA.

—— (1999b) *Towards a Regional Economic Development Strategy for the East of England: Background Papers*, Cambridge: EEDA.

Eckerberg, K. (1995) 'Environmental planning: dreams and reality', in A Khakee, I Elander and S Sunesson (eds) *Remaking the Welfare State: Swedish Urban Planning and Policy-making in the 1990s*, Aldershot: Avebury, 115–36.

Eckerberg, K. and Forsberg, B. (1998) 'Implementing Agenda 21 in local government: the Swedish experience', *Local Environment*, 3(3), 333–47.

Eisenschitz, A. and Gough, J. (1993) *The Politics of Local Economic Policy*, London: Macmillan.

Ekins, P. and Newby, L. (1998) 'Sustainable wealth creation at the local level in an age of globalisation', *Regional Studies*, 32(9), 863–71.

Elander, I. and Lidskog (2000) 'The Rio Declaration and subsequent global initiatives' in N. Low, B. Gleeson, I. Elander and R. Lidskog (eds) *Consuming Cities*, London: Routledge, 30–53.

Elkin, S. L. (1987) *City and Regime in the American Republic*, Chicago: University of Chicago Press.

Elkin, T., McLaren, D. and Hillman, M. (1991) *Reviving the City*, London: Friends of the Earth.

ENDS (1994) ''Screening' and 'scoping' proposed for environmental assessment', *ENDS Report*, 232, 38.

—— (1995) 'European Environment Agency gets under way', *ENDS Report*, 240, 20–3.

—— (1999) 'Environment on the agenda at international trade talks', *ENDS Report*, 295, 26–9.

Erdmenger, C. and Schreckenberger, S. (1998) *First Mover Advantage by Eco-efficiency – Local Incentives for Environment and Employment*, Freiburg: International Council for Local Environmental Initiatives.

European Commission (1994a) *Europe 2000+ Co-operation for European Territorial Development*, Luxembourg: Office for Official Publications of the European Communities.

—— (1994b) *Guide to the Community Initiatives 1994-99*, Luxembourg: Office for Official Publications of the European Communities.

—— (1995) *The Environment and the Regions: Towards Sustainability*, Brussels: European Commission.

REFERENCES

—— (1997) *Call for Tenders by Open Procedure No. 97.00.57.001 Concerning the Thematic Evaluation on the Impact of Structural Funds on the Environment*, Brussels: European Commission.

—— (1998a) *Sustainable Urban Development in the European Union: A Framework for Action*, Communication from the Commission to the Council, the European Parliament, the Economic and Social Committee and the Committee of the Regions, Brussels: DG XVI, European Commission.

—— (1998b) *Response of the EC Expert Group on the Urban Environment on the Communication 'Towards an Urban Agenda in the European Union'*, Brussels: DG XI, European Commission.

—— (1999) *Europe's Environment: What Directions for the Future? The Global Assessment of the European Community Programme of Policy and Action in Relation to the Environment and Sustainable Development, 'Towards Sustainability'*, Brussels: European Commission.

European Commission/Environment Agency (1997) *Encouraging Environmentally Sustainable Development Through European Structural Fund Programmes in British Regions*, Recommendations from European Commission seminar, Manchester, 27–28 October, Brussels: European Commission.

Ewing, S. (1996) 'Whose Landcare? Observations on the role of 'community' in the Australian Landcare programme', *Local Environment*, 1(3), 259–76.

Expert Group on the Urban Environment (1994) *European Sustainable Cities*, First Report, Sustainable Cities Project, XI/822/94-EN, Brussels: DGXI, European Commission.

—— (1996a) *European Sustainable Cities*, Brussels: DGXI, European Commission.

—— (1996b) *Targeted Summary of the European Sustainable Cities Report for Local Authorities*, Brussels: DGXI, European Commission.

Falk, R. (1992) 'American hegemony and the Japanese challenge' in G. D. Hook and M. A. Weiner (eds) *The Internationalisation of Japan*, London: Routledge, 32–60.

Far Eastern Economic Review (1991) 'Focus: Environment in Asia', *Far Eastern Economic Review*, 153(38), 35–57.

Feldman, T. and Jonas, A. (2000) 'Sage scrub revolution? Property rights, political fragmentation, and conservation planning in Southern California under the Federal Endangered Species Act', *Annals of the Association of American Geographers*, 90(2), 256–92.

Flynn, A. and Marsden, T. (1995) 'Guest editorial: rural change, regulation and sustainability', *Environment and Planning A*, 27, 1180–92.

Forum for the Future (1998a) *Local Economy Programme Strategy*, unpublished report.

—— (1998b) *Case Studies of Sustainable Local Economic Development*, London: Forum for the Future.

Fowke, R. and Prasad, D. K. (1996) 'Sustainable development, cities and local government', *Australian Planner*, 33(2), 61–6.

Frenkel, S. (1994) 'Old theories in new places? Environmental determinism and bioregionalism', *Professional Geographer*, 46(3), 289–95.

Friends of the Earth (1990) *Beyond Rhetoric: An Economic Framework for Environmental Policy Development in the 1990s*, London: Friends of the Earth.

—— (1997) 'Regional Development Agencies – an Opportunity Lost?', *Briefing Paper*, London: Friends of the Earth.

REFERENCES

Galtung, J. (1986) 'Towards a new economics: on the theory and practice of self-reliance' in P. Ekins (ed.) *The Living Economy: A New Economics in the Making*, London: Routledge, 98–109.

Gandy, M. (1997) 'The making of a regulatory crisis: restructuring New York City's water supply', *Transactions of the Institute of British Geographers*, 22, 338–58.

Gibbs, D. C. (1993) *The Green Local Economy*, Manchester: Centre for Local Economic Strategies.

—— (1996) 'Integrating sustainable development and economic restructuring: a role for regulation theory?' *Geoforum*, 27(1), 1–10.

—— (1997) 'Urban sustainability and economic development in the United Kingdom: exploring the contradictions', *Cities*, 14(4), 203–8.

—— (1998) 'Regional development agencies and sustainable development', *Regional Studies*, 32, 365–8.

Gibbs, D. C. and Longhurst, J. W. S. (1995) 'Sustainable development and environmental technology: a comparison of policy in Japan and the European Union', *The Environmentalist*, 15, 196–201.

Gibbs, D. C., Longhurst, J. W. S. and Braithwaite, C. (1996) 'Moving towards sustainable development? Integrating economic development and the environment in local authorities', *Journal of Environmental Planning and Management*, 39(3), 317–32.

—— (1998) '"Struggling with sustainability": weak and strong interpretations of sustainable development within local authority policy', *Environment and Planning A*, 30, 1351–65.

Giddens, A. (1990) *The Consequences of Modernity*, Cambridge: Polity Press.

Gillespie, A., Goddard, J. B., Hepworth, M. and Williams, H. (1989) 'Information and communications technology and regional development: an information economy perspective', *OECD STI Review*, 5, April, Paris: OECD, 84–111.

Glasson, J. (1995) 'Regional planning and the environment: time for a SEA change', *Urban Studies*, 32(4-5), 713–31.

Goldblatt, D. (1996) *Social Theory and the Environment*, Cambridge: Polity Press.

Goldsmith, Z. (2001) 'Mr. Nader goes to Washington', *The Ecologist*, 31, 1, 27–30.

Goodwin, M., Cloke, P. and Milbourne, P. (1995) 'Regulation theory and rural research: theorising contemporary rural change', *Environment and Planning A*, 27, 1245–60.

Gough, J. and Eisenschitz, A. (1996) 'The modernisation of Britain and local economic policy: promise and contradictions', *Environment and Planning D: Society and Space*, 14, 203–19.

Gouldson, A. and Murphy, J. (1996) 'Ecological modernisation and the European Union', *Geoforum*, 27, 11–21.

—— (1997) 'Ecological modernisation: restructuring industrial economies', in M. Jacob (ed.) *Greening the Millennium? The New Politics of the Environment*, Oxford: Blackwell, 74–86.

Graham, S. and Marvin, S. (1996) *Telecommunications and the City – Electronic Spaces, Urban Places*, Routledge: London.

Greater London Council (1985) *The London Industrial Strategy*, London: GLC.

—— (1986) *The London Financial Strategy*, London: GLC.

REFERENCES

Griefahn, M. (1994) 'Initiatives in Lower Saxony to link ecology to economy' in R. Socolow, C. Andrews, F. Berkhout and V. Thomas (eds) *Industrial Ecology and Global Change*, Cambridge: Cambridge University Press, 423–8.

Hajer, M. (1993) 'Discourse coalitions and the institutionalisation of practice: the case of acid rain in Great Britain' in F. Fischer and J. Forester (eds) *The Argumentative Turn in Policy Analysis and Planning*, Durham, NC: Duke University Press.

—— (1995) *The Politics of Environmental Discourse: Ecological Modernisation and the Policy Process*, Oxford: Oxford University Press.

Hall, T. and Hubbard, P. (1996) 'The entrepreneurial city: new urban politics, new urban geographies?', *Progress in Human Geography*, 20(2), 153–74.

—— (1997) *The Entrepreneurial City: Politics, Regime and Representation*, London: Wiley.

Hallstrom, L. (1999) 'Industry versus ecology: environment in the new Europe', *Futures*, 31, 25–38.

Hams, T. and Morphet, J. (1994) 'Agenda 21 and Towards Sustainability: The EU approach to Rio', *European Information Service*, 147, 3–7.

Hanf, K. (1996) 'Implementing international environmental policies' in A. Blowers and P. Glasbergen (eds) *Environmental policy in an International Context: Prospects for Environmental Change*, London: Arnold, 197–221.

Harvey, D. (1989) 'From managerialism to entrepreneurialism: the transformation in urban governance in late capitalism', *Geografiska Annaler*, 71, 3–17.

—— (1993) 'The nature of environment: the dialectics of social and environmental change', *Socialist Register*, London: Merlin Press, 1–51.

—— (1996) *Justice, Nature and the Geography of Difference*, Oxford: Blackwell.

Haughton, G. (1997) 'Developing sustainable urban development models', *Cities*, 14(4), 189–95.

Haughton, G. and Hunter, C. (1994) *Sustainable Cities*, London: Jessica Kingsley/Regional Studies Association.

Hawke, R. (1989) *Our Country, Our Future: Statement on the Environment*, Canberra: Australian Government Publishing Service.

Hawken, P. (1987) *The Ecology of Commerce*, New York: Harper Business.

Hay, C. (1994) 'Environmental security and state legitimacy' in M. O'Connor (ed.) *Is Capitalism Sustainable? Political Economy and the Politics of Ecology*, New York: Guilford Press, 217–31.

Hay, C. and Jessop, B. (1995) 'The governance of local economic development and the development of local economic governance: a strategic relational approach', Lancaster Regionalism Working Group Papers (Governance Series), 53 (Paper presented to American Political Science Association Annual Conference, 30 August–1 September).

Healey, P. (1995) 'Discourses of integration: making frameworks for democratic urban planning' in P. Healey, S. Cameron, S. Davoudi, S. Graham and A. Madani-Pour (eds) *Managing Cities: The New Urban Context*, Chichester: Wiley, 251–72.

Healey, P. and Shaw, T. (1994) 'Changing meanings of "environment" in the British planning system', *Transactions of the Institute of British Geographers*, 19, 425–38.

Heaton, G., Repetto, R. and Sobin, R. (1991) *Transforming Technology: An Agenda* for *Environmentally Sustainable Growth in the 21st Century*, Washington, DC: World Resources Institute.

Hepworth, M. (1990) 'Planning for the information city: the challenge and response', *Urban Studies*, 27, 537–58.

HM Government (1990) *This Common Inheritance*, Cm 1200, London: HMSO.

—— (1994) *Sustainable Development*, Cm 2426, London: HMSO.

Hines, C. (2000) *Localisation: A Global Manifesto*, London: Earthscan.

Hirst, P. and Thompson, G. (1996) *Globalisation in Question*, Cambridge: Polity Press.

Hirst, P. and Zeitlin, J. (1992) 'Flexible specialisation versus post-Fordism: theory, evidence and policy implications' in M. Storper and A. Scott (eds) *Pathways to Industrialisation and Regional Development*, London: Routledge, 70–115.

Huber, J. (1982) *Die verlorene Unschuld der Ökologie* [*The Lost Innocence of Ecology: New Technologies and Superindustrialized Development*], Frankfurt am Main: Fischer Verlag.

—— (1985) *Die Regenbogengesellschaft: Ökologie und Sozialpolitik* [*The Rainbow Society: Ecology and Social Politics*], Frankfurt am Main: Fisher Verlag.

Hudson, R. (1994) 'New production concepts, new production geographies? Reflections on change in the automobile industry', *Transactions of the Institute of British Geographers*, 19, 331–45.

Hunt, C. (1992) 'Goss *is* greener – but to Queensland labor growth is still more important than Ecologically Sustainable Development', *Social Alternatives*, 11(2), 35–8.

Institute for European Environmental Policy (1994) *The State of Reporting by the EC Commission in Fulfilment of Obligations Contained in EC Environmental Legislation*, London: IEEP.

Jackson, T. and Roberts, P. (1997) 'Greening the Fife economy: ecological modernisation as a pathway for local economic development', *Journal of Environmental Planning and Management*, 40(5), 615–29.

Jacobs, M. (1991) *The Green Economy*, London: Pluto Press.

—— (1999) *Environmental Modernisation: The New Labour Agenda*, Fabian Pamphlet 591, London: Fabian Society.

Jacobs, M. and Stott, M. (1992) 'Sustainable development and the local economy', *Local Economy*, 7(3), 261–72.

Jahn, T. (1991) 'Neue Grün(en) – Politik?', *Kommune*, 9(6), 54.

James, P. (1993) 'Japanese business and the environment', *Greener Management International*, 4, 27–33.

Jamison, A. and Baark, E. (1999) 'National shades of green: comparing the Swedish and Danish styles in ecological modernisation', *Environmental Values*, 8, 199–218.

Jänicke, M. (1985) *Preventive Environmental Policy as Ecological Modernisation and Structural Policy*, Berlin: Wissenschaftszentrum.

—— (1992) 'Conditions for environmental policy success: an international comparison' in M. Jachtenfuchs and M. Strubel (eds) *Environmental Policy in Europe: Assessment, Challenges and Perspectives*, Baden-Baden: Nomos Verlagsgesellschaft, 71–91.

—— (1997) 'The political system's capacity for environmental policy' in M. Jänicke and H. Weidner (eds) *National Environmental Policies: A Comparative Study of Capacity Building*, Berlin: Springer, 1–24.

REFERENCES

Jessop, B. (1990) 'Regulation theories in retrospect and prospect', *Economy and Society*, 19, 153–216.
—— (1994) 'Post-Fordism and the State' in A. Amin (ed.) *Post-Fordism: A Reader*, Oxford: Blackwell, 251–79.
—— (1995) 'The regulation approach, governance and post-Fordism: alternative perspectives on economic and political change?', *Economy and Society*, 24, 307–33.
—— (1997) 'A neo-Gramscian approach to the regulation of urban regimes: accumulation strategies, hegemonic projects and governance' in M. Lauria (ed.) *Reconstructing Urban Regime Theory*, Thousand Oaks, CA: Sage, 51–73.
Katz, C. (1998) 'Whose nature, whose culture? Private productions of space and the "preservation" of nature' in B. Braun and N. Castree (eds) *Remaking Reality: Nature at the Millennium*, London: Routledge, 46–63.
Keating, M. (1993) *The Earth Summit's Agenda for Change: A Plain Language Version of Agenda 21 and the Other Rio Agreements*, Geneva: Centre for Our Common Future.
Keil, R. (1995) 'The environmental problematic in world cities' in P. Knox and P. Taylor (eds) *World Cities in a World System*, Cambridge: Cambridge University Press, 280–97.
Keller A. (1997) 'Strategic environmental assessment of the European Structural Fund Objective One Programme for the Highlands and Islands of Scotland', *European Environment*, 7, 63–8.
Korten, D. (1995) *When Corporations Rule the World*, West Hartford, CT: Kumarian Press.
Lake, R. W. (2000) 'Contradictions at the local scale: local implementation of Agenda 21 in the USA' in N. Low, B. Gleeson, I. Elander and R. Lidskog (eds) *Consuming Cities*, London: Routledge, 70–90.
Lang, T. and Hines, C. (1993) *The New Protectionism*, London: Earthscan.
Lauria, M. (1997) *Reconstructing Urban Regime Theory*, Thousand Oaks, CA: Sage.
Leadbetter, C. (2000) *Mind Over Matter: Greening the New Economy*, London: Green Alliance.
Leadbetter, C. and Willis, R. (2001) 'Minds over matter', *Green Futures*, 27, 36–9.
Leborgne, D. and Lipietz, A. (1992) 'Conceptual fallacies and open questions on post-Fordism' in M. Storper and A. J. Scott (eds) *Pathways to Industrialisation and Regional Development*, London: Routledge, 332–48.
Lee, Kai N. (1993) *Compass and Gyroscope: Integrating Science and Politics for the Environment*, Washington, DC: Island Press.
Leff, E. (1995) *Green Production: Toward an Environmental Rationality*, New York: Guilford Press.
—— (1996) 'Marxism and the environmental question: from the critical theory of production to an environmental rationality for sustainable development' in T. Benton (ed.) *The Greening of Marxism*, New York: Guilford, 137–56.
Leonardi, R. (1995) 'Regional development in Italy: Social capital and the Mezzogiorno', *Oxford Review of Economic Policy*, 11(2), 165–79.
Lewis, M. W. (1992) *Green Delusions: An Environmentalist Critique of Radical Environmentalism*, Durham, NC and London: Duke University Press.

REFERENCES

Lidskog, R. and Elander, I. (2000) 'After Rio: environmental policies and urban planning in Sweden' in N. Low, B. Gleeson, I. Elander and R. Lidskog (eds) *Consuming Cities*, London: Routledge, 197–318.

Lipietz, A. (1992a) *Towards a New Economic Order: Postfordism, Ecology and Democracy*, Cambridge: Polity Press.

—— (1992b) 'The regulation approach and capitalist crisis: an alternative compromise for the 1990s' in M. Dunford and G. Kafkalas (eds) *Cities and Regions in the New Europe: The Global-Local Interplay and Spatial Development Strategies*, London: Belhaven, 309–34.

Lloyd P. and Meegan R. (1996) 'Contested governance: European exposure in the English regions', *European Planning Studies*, 4, 75–97.

Local Government Association (1999) *Sustainable Credit Unions: Guidance Notes for Local Authorities*, London: LGA.

Local Government Management Board (1993) *A Framework for Local Sustainability*, Luton: LGMB.

—— (1994) *Local Agenda 21: Principles and Process*, Luton: LGMB.

Long Island Progressive Coalition (1997) *Long Island 2020: A Greenprint for a Sustainable Long Island*, New York: LIPC.

Longhurst, J. W. S., Raper, D. W., Lee, D., Heath, B., Conlon, D. E. and King, H. (1993) 'Acid deposition: a select review 1852–1990, Part 2', *Fuel*, 72(10), 1363–80.

Low Choy, D. (1992) 'Conservation strategies for local authorities', *Common Ground: The Community Landscape*, proceedings from the Australian Institute of Landscape Architects National Conference, Brisbane, 21–22 August.

Low, N., Gleeson, B., Elander, I. and Lidskog, R. (2000) 'After Rio: urban environmental governance?' in N. Low, B. Gleeson, I. Elander and R. Lidskog (eds) *Consuming Cities*, London: Routledge, 281–307.

Lowe, E. and Warren, J. L. (1996) *The Source of Value: An Executive Briefing and a Sourcebook on Industrial Ecology*, Richland, WA: Pacific Northwest National Laboratory.

Luke, T. (1995) 'Sustainable development as a power/knowledge system: the problem of "governmentality"' in F. Fischer and M. Black (eds) *Greening Environmental Policy: The Politics of a Sustainable Future*, London: Paul Chapman, 21–32.

Luke, T. W. (2000) 'A rough road out of Rio: the right-wing reaction in the United States against global environmentalism' in N. Low, B. Gleeson, I. Elander and R. Lidskog (eds) *Consuming Cities*, London: Routledge, 54–69.

Lundqvist, L. J. (1997) 'The process of capacity-building – towards a new stage in Swedish environmental policy and management' in L. Mez and H. Weidner (eds) *Umweltpolitik und Staatsversagen*, Berlin: Edition Sigma, 332–7.

—— (2000) 'Capacity-building or social construction? Explaining Sweden's shift towards ecological modernisation', *Geoforum*, 31, 21–32.

McGinnis, M. V., Woolley, J. and Gamman, J. (1999) 'Bioregional conflict resolution: rebuilding community in watershed planning and organising', *Environmental Management*, 24(1), 1–12.

McGrew, A. (1993) 'The political dynamics of the 'new' environmentalism' in D. Smith (ed.) *Business and the Environment: Implications of the New Environmentalism*, London: Paul Chapman, 12–26.

McManus, P. (1994) 'Sustainability and international political economy: a lasting relationship or a one-night stand?', paper presented to the conference 'Struggling with Sustainability', Staffordshire University, 15 September.

McTaggart, W. D. (1993) 'Bioregionalism and regional geography: place, people and networks', *Canadian Geographer*, 37(4), 307–19.

Marsden, T., Murdoch, J., Lowe, P., Munton, R. and Flynn, A. (1993) *Constructing the Countryside*, London: UCL Press.

Marshall, T. (1998) 'The conditions for environmentally intelligent regional governance: reflections from Lower Saxony', *Journal of Environmental Planning and Management*, 41(4), 421–43.

Martin, P. and Halpin, D. (1999) 'Local governance of Australian rural environments: the development and performance of discursive institutions', paper presented to the Annual Meeting of the Association of American Geographers, Honolulu, Hawaii.

Martin, R. and Townroe, P. (1992) 'Changing trends and pressures in regional development' in P. Townroe and R. Martin (eds) *Regional Development in the 1990s: The British Isles in Transition*, London: Jessica Kingsley/Regional Studies Association.

Mayer, M. (1994) 'Post-Fordist city politics' in A. Amin (ed.) *Post-Fordism: A Reader*, Oxford: Blackwell, 316–37.

Mehra, M. (1997) *Towards Sustainable Development for Local Authorities: Approaches, Experiences and Sources*, Copenhagen: European Environment Agency.

Miller, K. (1996) 'Balancing the scales: guidelines for increasing biodiversity's chances through bioregional management' in R. Breckwoldt (ed.) *Approaches to Bioregional Planning*, Biodiversity Series, Paper No. 10, Biodiversity Group, Canberra: Department of the Environment, Sport and Territories, 9–20.

Mills, L. (1997) 'Working for better environmental performance in SMEs: policy and action in European cities', report of the project Greening the Local Economy – European Approaches, Berlin: PACTE.

Mol, A. (1994) 'Ecological modernisation of industrial society: three strategic elements', *International Social Science Journal*, 121, 347–61.

Mol, A. and Spaargaren, G. (1993) 'Environment, modernity and the risk-society: the apocalyptic horizon of environmental reform', *International Sociology*, 8(4), 431–59.

—— (2000) 'Ecological modernisation theory in debate: a review' in A Mol and D Sonnenfield (eds) *Ecological Modernisation Around the World: Perspectives and Critical Debates*, London: Frank Cass, 17–49.

Mol, A. and Sonnenfeld, D. (2000) *Ecological Modernisation Around the World: Perspectives and Critical Debates*, London: Frank Cass.

Moulaert, F. and Swyngedouw, E. (1992) 'Accumulation and organisation in computer and communications industries: a regulationist approach' in P. Cooke, F. Moulaert, E. Swyngedouw, O. Weinstein and P. Wells (eds) *Towards Global Localisation: The Computing and Telecommunications Industries in Britain and France*, London: UCL Press, 39–60.

Murphy, J. (1999) 'Towards an explanation of environmental policy in advanced industrial countries', unpublished PhD thesis, University of Hull.

Myerson, G. and Rydin, Y. (1994) '"Environment" and "planning": a tale of the mundane and the sublime', *Environment and Planning D: Society and Space*, 12, 437–52.

REFERENCES

National Science Foundation (2000) *Towards A Comprehensive Geographical Perspective on Urban Sustainability*, Final Report of the 1998 National Science Foundation Workshop on Urban Sustainability, Rutgers, New Jersey: Centre for Urban Policy Research.

Newby, L. (1999) 'Sustainable local economic development: a new agenda for action?' *Local Environment*, 4(1), 67–72.

Newby, L. and Bell, D. (1996) 'Leicester's lessons in local sustainability', *Town and Country Planning*, 65(4), 101–2.

Newton, P. (1992) 'The new urban infrastructure: telecommunications and the urban economy', *Urban Futures*, 5, 54–75.

Nijkamp, P. and Perrels, A. (1994) *Sustainable Cities in Europe*, London: Earthscan.

Nishimura, H. (1989) *How To Conquer Air Pollution: A Japanese Experience*, Amsterdam: Elsevier.

North East Lincolnshire Council (1998) *Sustainability in North East Lincolnshire: Current Trends in Issues Affecting Our Quality of Life*, Grimsby: NELC.

North West Development Agency (1999) *Draft Regional Economic Strategy*, Manchester: NWDA.

O'Connor, J. (1994) 'Is sustainable capitalism possible?' in M. O'Connor (ed.) *Is Capitalism Sustainable? Political Economy and the Politics of Ecology*, New York: Guilford Press, 152–75.

—— (1996) 'The second contradiction of capitalism' in T. Benton (ed.) *The Greening of Marxism*, New York: Guilford Press, 197–221.

O'Riordan, T. (1992) 'The environment' in P. Cloke (ed.) *Policy and Change in Thatcher's Britain*, Oxford: Pergamon, 297–324.

O'Riordan, T. and Jordan, A. (1995) 'The precautionary principle in contemporary environmental politics', *Environmental Values*, 4, 191–212.

O'Riordan, T. and Voisey, H. (1997) *Sustainable Development in Western Europe: Coming to Terms with Agenda 21*, London: Frank Cass.

Office of Technology Assessment (1994) *Saving Energy in US Transportation*, Washington DC: OTA.

Owens, S. (1994) 'Land, limits and sustainability: a conceptual framework and some dilemmas for the planning system', *Transactions of the Institute of British Geographers*, 19, 439–56.

Pacione, M. (1997) 'Local exchange trading systems as a response to the globalisation of capitalism', *Urban Studies*, 34(8), 1179–99.

Painter, J. (1991) 'Regulation theory and local government', *Local Government Studies*, November/December, 23–44.

Pearce, D., Turner, R. K., O'Riordan, T., Adger, N., Atkinson, G., Brisson, I., Brown, K., Dubourg, R., Fankhauser, S., Jordan, A., Maddison, D., Moran, D. and Powell, J. (1994) *Blueprint 3: Measuring Sustainable Development*, Earthscan: London.

Peck, J. and Miyamachi, Y. (1994) 'Regulating Japan? Regulation theory versus the Japanese experience,' *Environment and Planning D; Society and Space*, 12, 639–74.

Peck, J. and Tickell, A. (1992) 'Local modes of social regulation? Regulation theory, Thatcherism and uneven development', *Geoforum*, 23, 347–63.

—— (1994a) 'Searching for a new institutional fix: the *after*Fordist crisis and

REFERENCES

global-local disorder' in A. Amin (ed.) *Post-Fordism: A Reader*, Oxford: Blackwell, 280–315.

—— (1994b) 'Jungle law breaks out: neoliberalism and global-local disorder', *Area*, 26, 317–26.

—— (1995) 'The social regulation of uneven development: 'regulatory deficit', England's South East and the collapse of Thatcherism', *Environment and Planning A* 27: 15–40.

Pezzey, J. (1992) 'Sustainability: an interdisciplinary guide', *Environmental Values*, 1, 321–62.

Pincetl, S. (1999) 'The politics of influence: democracy and the growth machine in Orange County, USA, in A. E. G. Jonas and D. Wilson (eds) *The Urban Growth Machine: Critical Perspectives Two Decades Later*, Albany, NY: State University of New York Press, 195–211.

Piore, M. and Sabel, C. (1984) *The Second Industrial Divide: Possibilities for Prosperity*, New York: Basic Books.

Porritt, J. (1990) *Where on Earth are We Going?* London: BBC Books.

Porter, M. (1990) *The Competitive Advantage of Nations*, New York: Free Press.

Ramsay, M. (1996) *Community, Culture and Economic Development: The Social Roots of Local Action*. Albany, NY: State University of New York Press.

Ravetz, J. (1996) 'Towards a sustainable city region', *Town and Country Planning*, 65(5), 152–4.

Redclift, M. and Woodgate, G. (1994) 'Sociology and the environment: discordant discourse?' in M. Redclift and T. Benton (eds) *Social Theory and the Global Environment*, London: Routledge, 51–66.

Roberts, P. (1994) 'Environmental sustainability and business: recognising the problem and taking positive action' in C. C. Williams and G. Haughton (eds) *Perspectives Towards Sustainable Environmental Development*, Aldershot: Avebury, 37–53.

—— (1995) *Environmentally Sustainable Business: A Local and Regional Perspective*, London: Paul Chapman Publishing.

Roberts, P. and Gouldson, A. (2000) 'Retrospect and prospect: designing strategies for integrated economic development and environmental management' in A. Gouldson and P. Roberts (eds) *Integrating Environment and Economy: Strategies for Local and Regional Government*, London: Routledge, 257–69.

Roberts, P. and Jackson, T. (1999) 'Incorporating the environment into European regional programmes – evolution, progress and prospects', *Town and Country Planning*, 68(3), 85–8.

Robertson, J. (1986) 'What comes after full employment?' in P. Ekins (ed.) *The Living Economy: A New Economic in the Making*, London: Routledge, 85–96.

Robins, N. and Trisoglio, A. (1992) 'Restructuring industry for sustainable development' in J. Holmberg (ed.) *Policies for a Small Planet*, London: Earthscan, 157–94.

Roseland, M. (1992) *Towards Sustainable Communities*, Ottawa: National Round Table on the Environment and the Economy.

—— (1997) 'Dimensions of the eco-city', *Cities*, 14(4), 197–202.

Sale, K. (1974) 'Mother of all' in S. Kumar (ed.) *The Schumacher Lectures*, Volume 2, London: Abacus.

—— (1985) *Dwellers in the Land: The Bioregional Vision*, San Francisco: Sierra Club Books.
Schmidheiny, S. (1992) *Changing Course: A Global Business Perspective on Development and the Environment*, Cambridge, MA: MIT Press.
Schreurs, M. A. (1997) 'A political system's capacity for global environmental leadership: a case study of Japan' in L. Mez and H. Weidner (eds) *Umweltpolitik und Staatsversagen*, Berlin: Edition Sigma, 323–31.
Selman, P. (1996) *Local Sustainability: Managing and Planning Ecologically Sound Places*, London: Paul Chapman Publishing.
Short, J. R. (1999) 'Urban imagineers: boosterism and the representation of cities' in A. E. G. Jonas and D. Wilson (eds) *The Urban Growth Machine*, Albany: State University of New York Press, 37–54.
Sklair, L. (1994) 'Global sociology and global environmental change' in M. Redclift and T. Benton (eds) *Social Theory and the Global Environment*, London: Routledge, 205–27.
Snyder, G. (1990) *The Practice of the Wild*, New York: North Point Press.
Spaargaren, G. and Mol, A. (1992) 'Sociology, environment and modernity: ecological modernisation as a theory of social change', *Society and Natural Resources*, 5, 323–44.
State of California (1990) *The California Telecommuting Pilot Project: Final Report*, Sacramento: State of California.
Stevenson, F. and Ball, J. (1998) 'Sustainability and materiality: the bioregional and cultural challenges to evaluation', *Local Environment*, 3(2), 191–209.
Stewart, M. (1994) 'Between Whitehall and townhall: the realignment of urban regeneration policy in England', *Policy and Politics*, 22(2), 133–45.
Stoker, G. (1990) 'Regulation theory, local government and the transition from Fordism' in D. S. King and J. Pierre (eds) *Challenges to Local Government*, London: Sage, 242–64.
Stone, C. N. (1989) *Regime Politics: Governing Atlanta, 1946–1988*. Lawrence: University of Kansas Press.
Stren, R. (1992) 'Conclusion' in R. Stren, R. White and J. Whitney (eds) *Sustainable Cities: Urbanisation and the Environment in International Perspective*, Boulder, CO: Westview Press, 307–15.
Swyngedouw, E. (1997) 'Neither global nor local: "Glocalization" and the politics of scale' in K. Cox (ed.) *Spaces of Globalisation: Reasserting the Power of the Local*, New York: Guilford Press, 137–66.
Taylor, M., Bobe, M. and Leonard, S. (1995) 'The business enterprise, power networks and environmental change' in M. Taylor (ed.) *Environmental Change: Industry, Power and Policy*, Aldershot: Avebury, 57–81.
Terazono, E. (1994) 'Green legislation will test Tokyo's sincerity', *Financial Times*, 9 March.
Thomas, K. (1994) 'Planning for sustainable development: an exploration of the potential role of town planners and the planning system' in C. C. Williams and G. Haughton (eds) *Perspectives Towards Sustainable Environmental Development*, Avebury: Aldershot.
Tickell, A. and Peck, J. (1992) 'Accumulation, regulation and the geographies of post-Fordism: missing links in regulationist research', *Progress in Human Geography*, 16, 190–218.

Tilley, F. and Fuller, T. (2000) 'Foresighting methods and their role in researching small firms and sustainability', *Futures*, 32, 149–61.

Torgerson, D. (1995) 'The uncertain quest for sustainability: public discourse and the politics of environmentalism' in F. Fischer and M. Black (eds) *Greening Environmental Policy: The Politics of a Sustainable Future*, London: Paul Chapman, 3–20.

Turner, R. K. (1993) 'Sustainability: principles and practice' in R. K. Turner (ed.) *Sustainable Environmental Economics and Management: Principles and Practice*, London: Belhaven, 3–36.

Tzoumis, K., McMahon, M. and Munro, J. (1998) 'Reconciling economic and environmental interests in the US National Brownfields Program: The legacy of Superfund and the missing contribution of innovative technology', *Journal of Urban Technology*, 5(3), 61–77.

Ungerer, H. (1990) *Telecommunications in Europe*, Brussels: Commission of the European Communities.

United Nations Conference on Environment and Development (1992) *Agenda 21 – Action Plan for the Next Century*, Rio de Janeiro: UNCED.

United Nations (1993) *Agenda 21: the United Nations Programme of Action from Rio*, New York: United Nations.

Utsunomiya, F. and Hase, T. (2000) 'Japanese urban policy: challenges of the Rio Earth Summit' in N. Low, B. Gleeson, I. Elander and R. Lidskog (eds) *Consuming Cities*, London: Routledge, 131–52.

Voisey, H. and O'Riordan, T. (1997) 'Governing institutions for sustainable development: the United Kingdom's national level approach' in T. O'Riordan and H. Voisey (eds) *Sustainable Development in Western Europe: Coming to Terms with Agenda 21*, London: Frank Cass, 24–53.

Wackernagel, M. and Rees, W. (1996) *Our Ecological Footprint: Reducing Human Impact on the Earth*, Gabriola Island, BC: New Society Publishers.

Walker, L., Cocklin, C. and Le Heron, R. (2000) 'Regulating for environmental improvement in the New Zealand forestry sector', *Geoforum*, 31, 281–97.

Wallace, D. (1995) *Environmental Policy and Industrial Innovation: Strategies in Europe, the US and Japan*, London: Earthscan.

Ward, K. G. (1996) 'Rereading urban regime theory: a sympathetic critique', *Geoforum*, 27(4), 427–38.

—— (1997) 'Coalitions in urban regeneration: a regime approach', *Environment and Planning A*, 29, 1493–506.

Ward, S. (1996) 'Local government and the environment in the 1990s', *Contemporary Political Studies*, 1, 845–57.

Weale, A. (1992) *The New Politics of Pollution*, Manchester: Manchester University Press.

Webster, F. (1994) 'What information society?', *The Information Society*, 10, 1–23.

Weizsäcker E. U. von, Lovins, A. B. and Lovins, L. H. (1997) *Factor Four: Doubling Wealth – Halving Resource Use*, St Leonards, NSW: Allen and Unwin.

Welford, R. (1995) *Environmental Strategy and Sustainable Development: The Corporate Challenge for the 21st Century*, London: Routledge.

—— (1997) *Hijacking Environmentalism: Corporate Responses to Sustainable Development*, London: Earthscan.

Welford, R. and Gouldson, A. (1993) *Environmental Management and Business Strategy*, London: Pitman.

REFERENCES

White, R. and Whitney, J. (1992) 'Cities and the environment: an overview' in R. Stren, R. White and J. Whitney (eds) *Sustainable Cities: Urbanisation and the Environment in International Perspective*, Boulder, CO: Westview Press, 8–51.

Whitfield, M. and Hart, D. (2000) 'American perspectives on economic development and environmental management: changing the federal-local balance' in A. Gouldson and P. Roberts (eds) *Integrating Environment and Economy: Strategies for Local and Regional Government*, London: Routledge, 39–52.

Whittaker, S. (1997) 'Are Australian councils 'willing and able' to implement Local Agenda 21?', *Local Environment*, 2(3), 319–28.

Wilbanks, T. (1994) '"Sustainable development" in geographic perspective', *Annals of the Association of American Geographers*, 84(4), 541–56.

Wilkinson, D. (1997) 'Towards sustainability in the European Union? Steps within the European Commission towards integrating the environment into other European Union policy sectors' in T. O'Riordan and H. Voisey (eds) *Sustainable Development in Western Europe: Coming to Terms with Agenda 21*, London: Frank Cass, 153–73.

World Commission on Environment and Development (1987) *Our Common Future*, [Brundtland Report] Oxford University Press: Oxford.

World Development Movement (1997) *Briefing on the Multilateral Agreement on Investment*, London: WDM.

Wright, I. (1995) 'Implementation of sustainable development by Australian local governments', *Environmental and Planning Law Journal*, 12(1), 54–61.

Yorkshire Forward (1999) *The Regional Economic Strategy for Yorkshire and the Humber: A Consultation Document*, Leeds: Yorkshire Forward.

INDEX

Note: Page references in **bold** refer to Tables

accounting, environment and 98
Action Programmes 63
Advanced Natural Fuels 132
Agenda 2000 65
Agenda 21 (UN) 2, 52–3, 54, 55, 66, 74
Akeler Developments 128
Amsterdam, Treaty of 57
appropriate technologies 96–7
Association of British Credit Union Ltd 125
Australia, environmental legislation in 73–6
Australian Local Government Association 75

Basic Environment Law (1993) (Japan) 71
Berlin Ecological Renovation Programme 122
best practice 147
Biodiversity Convention 37
bio-indicators 117
bioregionalism 100–6
Blair, Tony 40, 79
Bonn Convention on Migrating Species 37
'boosterism' 42
Bradford Business and Environment Support Team (BEST) 111
Bradford Business Environment Forum 111
Bradford City Challenge 111
Bradford's Business Local Agenda 21 111

British Columbia Round Table on the Environment and the Economy 89
Brundtland Report 2, 31, 52, 54, 73, 82
buildings, energy-efficient 127–8, 137
Bush, George W. 40, 67, 69, 83, 142, 148
business, advice and support for 99, 111–13
Business Circles (Sweden) 110, 112–13
business clubs/networks 112
business co-operation and partnership 99
Business Eco Logic project (London Borough of Sutton) 111
business-to-business trading 115
Buy Local Net Bulletin Board 116
Buy Local project (NE Lincolnshire) 115–16

Calderdale and Kirklees Green Business Network 112
call centres 32, 33
capacity building and training 95
capacity to act 140
causal stories 43
central government localism 42
City Challenge 46
Clean Air Act (1970) (USA) 66
Clinton, Bill 40, 67
cognitive-informational framework 12
Communism 18
community enterprise 95–6
Community Support Frameworks (CSFs) 60

INDEX

consensual capacity 11
Convention on Biological Diversity 52
Convention on the International Trade in Endangered Species (CITES) 37
Cork Clean Technology Centre 115
corporate environmentalism 19
counterproductivity theorists 10
CREATE project (Merseyside, UK) 124
credit unions 125

decentralisation of industry 29
deep greens 6, 88
Delegation for Ecologically Sustainable Development (DESD) (Sweden) 80
demand management 90
Department of Trade and Industry (DTI) 79
derelict land, reusing 98
design standards 98
developed land, reusing 98
Directorates-General (DGs) 61
dispersal of industry 31
distance learning 32, 114
dual network approach 90-1
Dunkerque Industrial Environment Planning Scheme 117
Dyfi ecopark (Machynlleth) 132-3

Earth Council 52
Earth Summit (Rio) (1992) 3, 51, 52, 53-4, 55, 66, 67, 74, 78, 82-3
East of England Development Agency (EEDA) 119
Eco Centre (Jarrow) 128
eco-industrial parks (EIPs) 113, 128-33
eco-industrial revolution 53
Ecological Fund (Germany) 105
ecological modernisation 2, 7-12, 19, 141, 143-4, 145-6
  political programmes 9-10
  strong 8, 9
  weak 8, 9
ecological municipality 127
Ecological Sustainable Development Strategy for Australia 73-4
ecological switchover 7, 8, 10, 11
ecologically sustainable capitalism 140
Ecologically Sustainable Development (ESD) 73, 145
eco-management and audit scheme (EMAS) 118
economic-technological framework 12

economisation of nature 8
economy, investment in 99
ecosystems approach to industry 97
Ecotech Innovation and Business Park (Swaffham) 132
education initiatives 96
Employee Commute Operations (ECO) rules 32
Emscher Park (Germany) 114-15, 117
end-of-pipe measures 36
Energy 21 Renewable Energy Park (Ebley, Stroud) 132
English Partnerships 77
Environment Agency (Japan) 73
Environment City initiative (UK) 92, 113
Environmental Action Programmes (EAPs) (EU) 57-60, 64, 102, 146
environmental auditing and assessment 97
environmental capacity 11
Environmental Impact Assessment (EIA) Directive 60
Environmental Impact Assessment (EIA) Law (1997) (Japan) 70, 73
Environmental Improvement Programme (Berlin) 115
Environmental Information and Observation Network (EIONET) 57
environmental legislation
  Australia 73-6
  Japan 70-3
  Sweden 79-82
  UK 76-9
  USA 66-70
Environmental Protection Agency (EPA) 66, 67
environmental quality 100
Environmental Vehicle Systems 132
Ethyl Corporation 39
Europe 2000+ Co-operation for European Territorial Development 62-3
European Agricultural Guidance and Guarantee Fund (EAGGF-G) 60, 61
European Environment Agency (EEA) 57
European Environmental Science and Technology Park (Turin) 121-2, 123

169

INDEX

European Regional Development Fund (ERDF) 60, 61, 138
European Social Fund (ESF) 60, 61
European Union
  environmental policy 51, 55–65, 83
  problems in 63–4
  Fifth Environmental Action Programme 2
  Structural Funds 134
Evaluating Progress working group (USA) 68
externality effect 13, 14

Fascism 18
Federation of Swedish Industry (Industiföbundet) 82
Fife Structure Plan 118
Financial Instrument of Fisheries Guidance (FIFG) 60
Fordism 15–16 17, 22, 27, 28–30, 142
Forum for the Future's Local Economy programme 95
Framework Convention on Climate Change 52
free trade 33–9
Free Trade of the Americas Agreement (FTAA) 36
Friends of the Earth Scotland 115

General Agreement on Tariffs and Trade (GATT) 34, 37
geographical information system (GIS) 62
Global Tomorrow Coalition's Sustainable America programme 68
globalisation 33–9, 44–6
glocalisation 26, 43
good practice 100
Gothernburg Environment Project (GEP) 127
Green Aid Plan (Japan) 72
green architecture 122
Green City Denmark 120–1
green city network 120
green consumerism 19, 115–16
Green Lights programme 67
Green Tech Centre (Vesoul, France) 121
green workplaces 124

Hawke, Bob 73, 74
hollowing out of the nation-state 26

home delivery services 32
Home-Grown Economy project (St Paul, Minnesota) 116

import-substitution 100
Index of Sustainable Economic Welfare 93
indicators of environmental improvement 133–5
industrial ecology 97, 128–33
industrial partnerships, local government and 125–7
industrialisation, goals of 53–4
information and communication technologies 30–3, 113–14
information economy 30–1
innovative capacity 11
Innovative Economic Strategies for Sustainable Communities working group (USA) 68
Integrated Local Area Planning (ILAP) 75
Integrated System for Implementing Sustainability (ISIS) 62
inter-generational equity 88
Inter-governmental Agreement on the Environment (IGAE) 74
Internal Environmental Care System (The Netherlands) 111
International Council for Local Environmental Initiatives (ICLEI) 67
inward investment, sustainable approaches to 98

Japan, environmental legislation 70–3
just-in-time production systems 33

Kalundborg eco-industrial park 129–30, 132
Keating, Paul 74
Kent Prospects Sustainable Business Partnership (SBP) 118
Keynesian Welfare State 27
Keynesianism 21
KONVER 60
Kyoto agreement (1997) 1, 66, 69, 142, 148

land use 98, 110
Landcare 75
law and insurance 98
LEADER 60

INDEX

Leicester's Environment City 92
LIFE 61, 62
Local Agenda 21 (LA21) 53, 55, 74, 75, 92, 103, 118
  Sweden 81–2, 124
  UK 77, 78, 79
  USA 67
local economic development 86–7
Local Environmental Action Plans (LEAPs) 135
Local Exchange Trading Schemes (LETS) 49, 78, 125
local government partnerships, industry and 125–7
local labour markets 122–5
London Industrial and Financial Strategies (GLC) 41
Lothian and Edinburgh Environmental Partnership (LEEP) 120, 125

Maastricht Treaty (1992) 55–6, 64
Mackenzie Basin Impact Study 102
marketing and public relations 98
Medicaid 39
Medicare 39
Meet the Buyer evenings 116
Minamata mercury disaster 70
Ministry of International Trade and Industry (MITI) (Japan) 71
Minnesota 'Milestones' programme 69
  Sustainable Development Initiative 69
Mitsubishi Kasei 74
MMT 39
modes of social regulation (MSR) 21–2, 25
Multi-jurisdictional and Regional Collaboration working group (USA) 68
Multilateral Agreement on Investment (MAI) 27, 38–9

National Agenda 21 for Japan 71
National Brownfields Program 67
National Environmental Policy Plan (The Netherlands) 105
National Sustainability Program (Australia) 75–6
neo-liberalism 18–19, 39–43
New Deal initiative 137
New Earth 21 (Japan) 72
new economic spaces 43
New Economics Foundation 134

new localism 43, 44
New Sunshine Programme (Japan) 72
NIMBYism 13
North American Agreement on Environmental Co-operation 35
North American Free Trade Agreement (NAFTA) 34, 35–6, 39, 45

Oregon Department of Environmental Quality 32
Oregon's 'Benchmarks' programme 69
Organisation for Economic Co-operation and Development (OECD) 38
Our Common Future 52
Our Country, Our Future 73

packaging and product design 98
Pax Britannica 21
performance partnerships 42
political-institutional framework 12
polluter pays principle 90
Portes des Alpes technology park (Lyon, France) 121
post-Fordism 21, 27, 28–30, 41, 102, 142
  optimistic 28
  pessimistic 28
precautionary principle 89
President's Council on Sustainable Development (PCSD) (USA) 67–8, 69. 130, 131
Principles of Forestry Management 52

Queensland's Environmental Protection Act (1994) 73

Reaganomics 39
RECHAR 60
regional conversion plans 102
Regional Development Agencies (RDAs) 77, 105, 119, 136
regional environmental management systems 100–6, 117
REGIS 60
regulation theory 2, 15–23, 25, 141–2, 144
  modes of regulation 18
  phases of regulation 15–16, **16**
Research Institute for the Earth (RITE) (Japan) 71

171

INDEX

RETEX 60, 111
Rio Plus Ten summit (2002) 55
Rogernomics 40
Round Table on Sustainable Development (UK) 76, 78
Rural Development Commission 77
Rural Strategies working group (USA) 68

Sala Eco centre 124
Schumpeterian Workfare State 27
sector studies 100
self-reliant cities 47
Single European Market 23, 64
Single Programming Documents (SPDs) 64
Single Regeneration Budget (UK) 46–7, 138
sites of intervention 148
situative context 140
small and medium-sized firms (SMEs) 92, 99, 111, 115–16
social democracy 18
social justice 88
social regulation, modes of (MSR) 21–2, 25
social welfare provision 39, 40
Solar Building (Doxford business park) 127–8
space of flows 46
strategic capacity 11
Strategic Environmental Assessments 60
Structural Fund Regulations 60, 61, 65
Surface Mining Control and Reclamation Act (1977) (USA) 66
sustainability appraisal 133–5
Sustainability Indicators Research Project (UK Local Government Management Board) 134
sustainability infrastructure 127–8
Sustainable America 69
sustainable and fulfilling work 96
sustainable business park 116
sustainable city 5
sustainable development 2, 3–7, 19, 24–5, 48–9, 78
  criticisms of 5–7
  ecological principles 89–90
  governance principles 91–2
  impact principles 93
  principles in local areas and regions 88–93

resource efficiency principles 90–1
  spectrum of 3–4, **4**, **5**
  strong versions 3, 4–5, 7
  weak versions 3, 5, 19, 21–2
Sustainable Development: The UK's Strategy 76
sustainable employment 96
sustainable local and regional economies, principles for 93–100
  business themes 96–8
  community and work-based themes 95–6
  themes for local and regional government 98–100
sustainable local economic development (SLED) 95
Sustainable Production Programme (Bologna) 126
'Sustainable Sweden, A' 80
Sweden, environmental legislation 79–82
Swedish Eco-Municipality Network 81

Takeshita, Prime Minister 71
targeted inward investment strategy 116–17
Tasmania's Resource Management Planning System 73
Technology Renaissance for the Environment and Energy (TREE) (Japan) 73
technology, using for sustainability 113–15
telebanking 32, 114
telecommuting/teleworking 32, 114
teleconferencing 32, 114
telecottages 114
telematics 114
telemedicine 32, 114
telephone booking 32
teleservices 114
teleworking *see* telecommuting
Thatcher, Margaret 41
Thatcherism 39
Third Way 40
This Common Inheritance 76–7
Tobin tax 39
training initiatives 96
transaction cost approaches 13–14
transfrontier justice 88
transport policies 99–100, 110

INDEX

UK, environmental legislation 76–9
United Nations
  Agenda 21 programme 2, 52–3, 54, 55, 66, 74
  Commission on Environment and Development Earth Summit (Rio de Janeiro) (1992) 3, 51, 52, 55, 66, 74, 78, 82–3
  Commission on Sustainable Development 52
  Habitat II conference (City Summit) 54
  World Commission on Environment and Development (WCED) (Brundtland Report) (UN) 2, 31, 52, 54, 73, 82
URBAN 60
urban growth coalitions 42
urban regime theory 2, 12–15, 25, 144
USA, environmental legislation 66–70

Victoria's Sustainable Development Program for 1995–2000 73
voluntary agreements 99

Vreten Business Circle (Sweden) 110, 112

waste exchange networks 115
Water Pollution Control Act (1972) (USA) 66
waystations 135
Welsh Assembly 77
Wiltshire Centre for Sustainable Development 111
wired communities 114
Wise Group 122
Work Instead of Income Support 124
World Commission on Environment and Development (WCED) (Brundtland Report) UN) 2, 31, 52, 54, 73, 82
World Trade Organisation (WTO) 27, 34, 37, 38, 41
  Committee on Trade and the Environment 39

ZAUG GmbH 124
zero-energy office 128